Texts and Readings in Mathematics

Volume 86

Managing Editor
Rajendra Bhatia, Ashoka University, Sonepat, Haryana, India

Editorial Board
Manindra Agrawal, Indian Institute of Technology, Kanpur, India
V. Balaji, Chennai Mathematical Institute, Chennai, India
R. B. Bapat, Indian Statistical Institute, New Delhi, India
V. S. Borkar, Indian Institute of Technology, Mumbai, India
Apoorva Khare, Indian Institute of Science, Bangalore, India
T. R. Ramadas, Chennai Mathematical Institute, Chennai, India
V. Srinivas, Tata Institute of Fundamental Research, Mumbai, India

The **Texts and Readings in Mathematics** series publishes high-quality textbooks, research-level monographs, lecture notes and contributed volumes. Undergraduate and graduate students of mathematics, research scholars and teachers would find this book series useful. The volumes are carefully written as teaching aids and highlight characteristic features of the theory. Books in this series are co-published with Hindustan Book Agency, New Delhi, India.

Arup Bose · Arijit Chakrabarty ·
Rajat Subhra Hazra

A Little Book of Martingales

Arup Bose
Statistics and Mathematics Unit
Indian Statistical Institute
Kolkata, West Bengal, India

Arijit Chakrabarty
Statistics and Mathematics Unit
Indian Statistical Institute
Kolkata, West Bengal, India

Rajat Subhra Hazra
Mathematical Institute
Leiden University
Leiden, The Netherlands

ISSN 2366-8717   ISSN 2366-8725 (electronic)
Texts and Readings in Mathematics
ISBN 978-981-97-4471-8   ISBN 978-981-97-4472-5 (eBook)
https://doi.org/10.1007/978-981-97-4472-5

Jointly published with Hindustan Book Agency
The print edition is not for sale in India. Customers from India please order the print book from: Hindustan Book Agency, P 19 Green Park Extension, New Delhi 110016, India.
ISBN of the Co-Publisher's edition: 978-81-957829-6-3

Mathematics Subject Classification: 60G42, 60F20, 60G46, 60-01, 60F05, 60B10

© Hindustan Book Agency 2024

This work is subject to copyright. All rights are solely and exclusively licensed by the Publisher, whether the whole or part of the material is concerned, specifically the rights of translation, reprinting, reuse of illustrations, recitation, broadcasting, reproduction on microfilms or in any other physical way, and transmission or information storage and retrieval, electronic adaptation, computer software, or by similar or dissimilar methodology now known or hereafter developed.
The use of general descriptive names, registered names, trademarks, service marks, etc. in this publication does not imply, even in the absence of a specific statement, that such names are exempt from the relevant protective laws and regulations and therefore free for general use.
The publisher, the authors and the editors are safe to assume that the advice and information in this book are believed to be true and accurate at the date of publication. Neither the publisher nor the authors or the editors give a warranty, expressed or implied, with respect to the material contained herein or for any errors or omissions that may have been made. The publisher remains neutral with regard to jurisdictional claims in published maps and institutional affiliations.

This Springer imprint is published by the registered company Springer Nature Singapore Pte Ltd.
The registered company address is: 152 Beach Road, #21-01/04 Gateway East, Singapore 189721, Singapore

If disposing of this product, please recycle the paper.

*In memory of
Saurabh Ghosh*

Preface

Martingales have applications in statistics, mathematics, economics, and other disciplines. Discrete time martingales have been treated in numerous books and research articles. Mention could be made of Billingsley (1995), Breiman (1968), Chung (2001), Durrett (2019), Chow and Teicher (2003), Doob (1990), Hall and Heyde (1980), Neveu (1975), and Williams (1991). This little book covers the basic results on discrete time martingales and provides some application of these results. We have of course drawn generously from the above books and other materials available.

The authors have taught Masters level courses on martingales in discrete time. This book has grown out of that. It covers the syllabus for these courses in the M.Stat. and M.Math programs of the Indian Statistical Institute. We have added quite a few additional topics and results that we feel are interesting and useful. These would provide the instructor liberty to move beyond the syllabus should the students be up for it. Adequate details are provided for the benefit of the students, along with exercises within the text and at the end of the chapters. Additional references are provided for the students to explore further.

An exposure to a first course in measure theory is a prerequisite. For convenience, in Chap. 1 we have recollected the essential concepts and results, mostly without proofs. Readers who need more details can refer to one of many well-known books available, for instance, Durrett (2019) or Billingsley (1995). We have used mostly Ash and Doléans-Dade (2000) for this and the next chapter.

Chapter 2 is on signed measures. The Jordan–Hahn theorem decomposes a signed measure into the difference of two measures in a unique way. This result is crucial to prove the Radon–Nikodym theorem which generalizes the concepts of the derivative and the indefinite integral, to measures. It leads to the Lebesgue decomposition theorem.

In Chap. 3, we introduce the notion of conditional expectation which is based on the Radon–Nikodym theorem. We establish its basic properties, including Jensen's inequality. We also show the existence of a regular conditional probability distribution. The useful notion of uniform integrability is also a part of this chapter.

We introduce martingales, sub- and super-martingales in Chap. 4. We define stopping times, which is fundamental in martingale theory, and prove Doob's Optional

Sampling Theorem. This result essentially says that no legitimate strategy can alter the nature (that is, from fair to one sided) of a two-person game. The famous identities of Wald follow from this theorem. Doob's maximal inequality shows how the maximum of a sub-martingale is controlled by the last point of the sequence probabilistically.

Chapter 5 is on convergence in L^p and almost sure convergence results for martingales, sub-martingales and super-martingales. The crucial tool in developing these results is the Upcrossing lemma. It gives a bound for the expectation of the number of upcrossings of any interval by a sub-martingale, in terms of the expectation of the end point of the sequence and the end points of the interval. There are numerous results on the convergence of martingales, and we provide a sampling of these. The notion of reverse martingale (time-reversed martingales) is also introduced in this chapter.

Chapter 6 gives some applications of the results from Chap. 5. We show how Kolmogorov 0-1 law and Hewitt–Savage 0-1 law follow from the reverse martingale convergence theorem. The strong law of large numbers for average of independent and identically random variables, as well as for U-statistics, are proved by using reverse martingales. The strong law for exchangeable sequences is also established. We also state and prove de-Finetti's theorem for exchangeable random variables. The final topic in this chapter is Kakutani's theorem for product martingales, which has application in statistics.

Chapter 7 is on the central limit theorem (CLT). CLT for martingales is very useful, since it can be used in numerous dependent models. We state and prove one of the simplest versions of CLT for martingales. As simple illustrations, we apply it to a simple urn model, to the trace of a random matrix, and to Markov chains.

In Chap. 8 we cover some additional topics such as forward martingale representation for U-statistics, extended/conditional Borel–Cantelli lemma, Azuma–Hoeffding inequality, conditional three series theorem, strong law for martingales, and the Kesten–Stigum theorem for a simple branching process. The so-called Burkholder inequalities are also covered in this chapter.

We thank Bishakh Bhattacharya and Aditya Guha Roy for proof reading parts of the book. Valuable comments were provided by Ayan Bhattacharya, who in particular provided the material on the Kesten–Stigum theorem. We also thank Aniket Jain for helpful discussions on alternate proofs of de Finetti's theorem.

Finally, we thank Rajendra Bhatia for encouraging us to write in the TRIM series, and to the Referees for providing encouraging and constructive comments on our draft in lightning quick time.

The work of Arup Bose has been partially supported by the J. C. Bose Fellowship JBR/2023/000023 from the Science and Education Research Board, Government of India.

Kolkata, India	Arup Bose
Kolkata, India	Arijit Chakrabarty
Leiden, The Netherlands	Rajat Subhra Hazra
February 2024	

References

R. Ash, C. A. Doléans-Dade, *Probablility & Measure Theory* (Elsevier Science, 2000)

P. Billingsley, *Probability and Measure* (John Wiley, 1995)

L. Breiman, *Probability* (Addison-Wesley Publishing Co., Reading, Mass.-London-Don Mills, Ont., 1968)

Y. S. Chow, H. Teicher, *Probability Theory: Independence, Interchangeability, Martingales* (Springer Science & Business Media, 2003)

K. L. Chung. *A Course in Probability Theory*, 3rd edn. (Academic Press, Inc., San Diego, CA, 2001)

J. L. Doob, *Stochastic Processes. Wiley Classics Library* (Wiley-Interscience, 1990)

R. Durrett. *Probability: Theory and Examples*, vol. 49 (Cambridge University Press, 2019)

P. Hall, C. C. Heyde, *Martingale Limit Theory and Its Application* (Academic Press, Inc. Harcourt Brace Jovanovich, Publishers, New York-London, 1980)

J. Neveu, *Discrete-parameter Martingales* (North-Holland Publishing Co., Amsterdam, 1975)

D. Williams, *Probability with Martingales* (Cambridge Mathematical Textbooks. Cambridge University Press, Cambridge, 1991)

Contents

1 **Measure** 1
   1.1   Monotone Class Theorem 1
   1.2   Borel σ-Field 2
   1.3   Measure 3
   1.4   Measurable Function and Integration 4
   1.5   Convergence 6
   1.6 L^p Space 6
   1.7   Distribution Function 8
   1.8   Lebesgue-Stieltjes Measure 8
   1.9   Lebesgue Measure 12
   1.10 Probability Distribution 14
   1.11 Characteristic Function 17
   1.12 Independence 18
   1.13 Exercises 19
   References 21

2 **Signed Measure** 23
   2.1   Jordan–Hahn Decomposition 23
   2.2   Absolute Continuity, Radon–Nikodym Theorem 27
   2.3   Singularity and Lebesgue Decomposition 32
   2.4   Absolutely Continuous Function 34
   2.5   Exercises 36
   Reference 38

3 **Conditional Expectation** 39
   3.1   Conditional Expectation 39
   3.2   Regular Conditional Distribution 48
   3.3   Jensen's Inequality 50
   3.4   Uniform Integrability 52
   3.5   Exercises 57
   Reference 60

4 Martingales ... 61
- 4.1 Definition and Examples ... 61
- 4.2 Stopping Time ... 67
- 4.3 Doob's Optional Stopping Theorem ... 70
- 4.4 Wald's Identities ... 76
- 4.5 Maximal Inequality ... 78
- 4.6 `Abracadabra` ... 80
- 4.7 Exercises ... 83
- References ... 84

5 Almost Sure and L^p Convergence ... 85
- 5.1 Upcrossing Lemma ... 85
- 5.2 L^1-Bounded and UI Sub-martingales ... 87
- 5.3 L^p Convergence for $p > 1$ ... 89
- 5.4 Reverse Martingale ... 91
- 5.5 Exercises ... 93
- References ... 96

6 Application of Convergence Theorems ... 97
- 6.1 Kolmogorov's 0-1 Law ... 97
- 6.2 Strong Law for iid Sequences ... 98
- 6.3 Strong Law for U-Statistics ... 100
- 6.4 Hewitt–Savage 0-1 Law ... 103
- 6.5 Strong Law for Exchangeable Sequences ... 105
- 6.6 de Finetti's Theorem ... 107
- 6.7 Kakutani's Theorem for Martingales ... 111
- 6.8 Likelihood Ratio ... 113
- 6.9 Exercises ... 115
- References ... 118

7 Central Limit Theorem ... 119
- 7.1 Central Limit Theorem: Independent Summands ... 119
- 7.2 Martingale Central Limit Theorem ... 120
- 7.3 Urn Model ... 126
- 7.4 Random Matrix ... 129
- 7.5 Markov Chains ... 133
- References ... 139

8 Additional Topics ... 141
- 8.1 U-Statistics, Forward Martingale Representation ... 141
- 8.2 Azuma–Hoeffding Inequality ... 143
- 8.3 Weak Law for Martingales ... 148
- 8.4 Extended Borel–Cantelli Lemma ... 149
- 8.5 Three-Series Theorem ... 151
- 8.6 Toeplitz and Kronecker Lemma ... 152

8.7	Strong Law of Large Numbers for Martingales	154
8.8	Burkholder–Davis–Gundy Inequalities	157
8.9	Branching Process: Kesten–Stigum Theorem	167
References		173

Index 175

Author Index 181

Notation and Abbreviations

$\mathbb{R}$ and $\mathbb{R}^+$: set of, real, and non-negative real numbers.
$\bar{\mathbb{R}} = \mathbb{R} \cup \{\infty\} \cup \{-\infty\}$.
$\mathbb{Q}$: set of rational numbers.
$\mathbb{N}$: set of positive integers $\{1, 2, \cdots\}$, sometimes with 0 included.
$a \wedge b$: minimum of a and b.
$A \cup B$ and $A \cap B$: union and intersection of the sets A and B.
$\limsup A_n = \bigcap_{n=1}^{\infty} \bigcup_{k=n}^{\infty} A_k$, $\liminf A_n = \bigcup_{n=1}^{\infty} \bigcap_{k=n}^{\infty} A_k$.
$\mathbf{1}_A$: indicator function of the set A.
$\sigma(\mathcal{C})$: smallest σ-field containing the class of sets $\mathcal{C}$.
$\sigma(\mathcal{X})$: smallest σ-field that makes all variables in the class $\mathcal{X}$ measurable.
$\mathcal{B}(M)$: smallest σ-field containing all open sets of M (the Borel σ-field).
$\mathcal{G} \vee \mathcal{H}$: smallest σ-field containing $\mathcal{G}$ and $\mathcal{H}$.
$(\Omega, \mathcal{A}, P)$: probability space.
E, Var, Cov: expectation, variance, and covariance.
$X \stackrel{d}{=} Y$: X and Y have the same probability distribution.
$X_n \Rightarrow X$: X_n converges to X in distribution.
$F_n \Rightarrow F$: Probability distributions F_n converges weakly to F.
$X_n \stackrel{P}{\to} X$: X_n converges to X in probability.
DCT: dominated convergence theorem.
OST: Doob's optional stopping theorem.
UI: uniformly integrable.
a.s.: almost surely.
iid: independent and identically distributed.

Equations are numbered within a chapter. Definitions, theorems, examples, remarks, lemmas, corollaries are numbered within a section, within a chapter.

Chapter 1
Measure

This chapter recalls the basic notions of measure and integration that will be essential for us. Most of the results are stated without proofs. We recall the notions of semi-field, field, and σ-field. We state the monotone class theorem and Carathéodory's extension theorem. Concepts of independence, convergence in measure/probability, almost surely convergence, and L^p-spaces are recalled. We state the different convergence results such as the Borel–Cantelli lemma, Fatou's lemma, the monotone convergence theorem, the dominated convergence theorem, and Fubini's theorem. Construction of the Lebesgue–Stieltjes measures is given in some details. We recall Lévy's continuity theorem for characteristic functions. For more details, see Breiman (1968) or Billingsley (1995). We have used mostly Ash and Doléans-Dade (2000) for this and the next chapter.

1.1 Monotone Class Theorem

Suppose Ω is a non-empty set.

Definition 1.1.1 (*Semi-field*) Any non-empty collection $\mathcal{S}$ of subsets of Ω is called a **semi-field** if

(a) $\Omega \in \mathcal{S}$,

(b) it is closed under finite intersection,

(c) for any $A \in \mathcal{S}$, A^c is a finite disjoint union of sets from $\mathcal{S}$. ◇

Definition 1.1.2 (*Field*) Any non-empty collection $\mathcal{F}$ of subsets of Ω is called a **field** or an **algebra** if

(a) $\Omega \in \mathcal{F}$,

(b) $A \in \mathcal{F}$ implies $A^c \in \mathcal{F}$,

(c) $\{A_i\}_{1 \leq i \leq n}$ belong to $\mathcal{F}$ implies $\cup_{i=1}^n A_i \in \mathcal{F}$. ◇

It can be checked that, without any loss, in Condition (c) intersection can be replaced by union. Moreover, any field is a semi-field.

Definition 1.1.3 (σ-*field*) A non-empty class $\mathcal{A}$ of subsets of Ω is called a σ-**field** or a σ-**algebra** if $\Omega \in \mathcal{G}$, and is closed under the operations of complementation and countable union. Then $(\Omega, \mathcal{A})$ is called a **measurable space**. ◇

Thus any σ-field is a field. Arbitrary intersection of σ-fields (of subsets of Ω) is a σ-field. Hence, for any collection of subsets $\mathcal{C}$ of Ω, the smallest σ-field that contains $\mathcal{C}$ exists, and shall be denoted by $\sigma(\mathcal{C})$. If $\{\mathcal{G}_\alpha\}_{\alpha \in I}$ is a collection of σ-fields of subsets of Ω, then $\cup_{\alpha \in I} \mathcal{G}_\alpha$ need not be a σ-field. The smallest σ-field containing $\cup_{\alpha \in I} \mathcal{G}_\alpha$ will be denoted by

$$\bigvee_{\alpha \in I} \mathcal{G}_\alpha := \sigma\big(\cup_{\alpha \in I} \mathcal{G}_\alpha\big).$$

The concept of a monotone class will be crucial for us.

Definition 1.1.4 (*Monotone class*) Suppose Ω is a non-empty set. Any non-empty collection $\mathcal{M}$ of subsets of Ω is called a **monotone class** if it is closed under both increasing and decreasing limits of sets. ◇

Exercise 1.1.1 (a) Show that arbitrary intersection of fields is a field.

(b) Show that arbitrary intersection of monotone classes is a monotone class.

We recall the Monotone class theorem without proof. This result is not hard to prove but serves as a very crucial tool.

Theorem 1.1.1 (Monotone class theorem) *Suppose Ω is a non-empty set. Suppose $\mathcal{F}$ is a field, $\mathcal{M}$ is a monotone class, and $\mathcal{M} \supset \mathcal{F}$. Then $\mathcal{M} \supset \sigma(\mathcal{F})$.* ◆

1.2 Borel σ-Field

The sets $\mathbb{R}, \mathbb{R}^+$, and $\bar{\mathbb{R}} = \mathbb{R} \cup \{\infty\} \cup \{-\infty\}$ denote all real numbers, all non-negative real numbers, and all extended real numbers, respectively. The arithmetic when $\pm\infty$ and 0 are involved stipulate that $\infty - \infty$ and $-\infty + \infty$ are meaningless, $\infty + \infty = \infty$, $-\infty - \infty = -\infty$, $\infty \times 0 = -\infty \times 0 = 0$, and $0^0 = 1$.

Given any topological space M, its Borel σ-field is the smallest σ-field containing all open sets. It is denoted by $\mathcal{B}(M)$ or simply by $\mathcal{B}$ if M is clear from the context. The elements of a Borel σ-field are called **Borel sets**. Thus $\mathcal{B}(\mathbb{R})$ is the smallest σ-field containing all open intervals, and $\mathcal{B}(\mathbb{R}^N)$ is the smallest σ-field containing all rectangles of $\mathbb{R}^N$. The Borel σ-field for $\mathbb{R}^+$, $\bar{\mathbb{R}}$, and their n-fold products are defined similarly.

Exercise 1.2.1 (a) Let $\mathcal{S}$ be a semi-field, and $\mathcal{F}$ consists of all sets which are finite disjoint union of sets from $\mathcal{S}$. Show that $\mathcal{F}$ is a field.
(b) Define

$$\bar{\mathcal{S}} := \{(a, b] \cap \mathbb{R} : a, b \in \bar{\mathbb{R}}, \ a \leq b\},$$
$$\bar{\mathcal{F}} := \{\cup_{k=1}^{n} I_k : I_k \in \bar{\mathcal{S}} \text{ for } k = 1, \ldots, n \text{ and are disjoint}\}.$$

Show that $\bar{\mathcal{S}}$ and $\bar{\mathcal{F}}$ are, respectively, semi-field and field of subsets of $\mathbb{R}$.

1.3 Measure

Let $\mathcal{F}$ be a field of subsets of Ω.

Definition 1.3.1 (*Finite and countable additivity*) (a) A set function $\mu : \mathcal{F} \to \mathbb{R}^+ \cup \{\infty\}$ is said to be **finitely additive** if

$$\mu(\bigcup_{i=1}^{n} A_i) = \sum_{i=1}^{n} \mu(A_i), \text{ for all disjoint } \{A_i\}_{1 \leq i \leq n} \text{ from } \mathcal{F}.$$

(b) A function $\mu : \mathcal{F} \to \mathbb{R}^+ \cup \{\infty\}$ is said to be a **measure** on $\mathcal{F}$, if it is **countably additive**: for all disjoint $\{A_i\}$ from $\mathcal{F}$ with $\cup_{i=1}^{\infty} A_i \in \mathcal{F}$,

$$\mu(\bigcup_{i=1}^{\infty} A_i) = \sum_{i=1}^{\infty} \mu(A_i).$$

◇

Exercise 1.3.1 (*Countable sub-additivity*) Let μ be a measure on a field $\mathcal{F}$. Let $\{A_i\}$ be a sequence from $\mathcal{F}$ which are not necessarily disjoint, such that $\cup_{i=1}^{\infty} A_i \in \mathcal{F}$. Show that

$$\mu(\bigcup_{i=1}^{\infty} A_i) \leq \sum_{i=1}^{\infty} \mu(A_i).$$

We usually work with measures on σ-fields. If $\mathcal{A}$ is a σ-field on Ω and μ a measure on $(\Omega, \mathcal{A})$, then $(\Omega, \mathcal{A}, \mu)$ is a measure space. If $\mu(\Omega) < \infty$, then μ is called a **finite measure**. If there are sets $\{A_i\}_{i \geq 1}$ from $\mathcal{F}$ (or $\mathcal{A}$) such that $\cup_{i=1}^{\infty} A_i = \Omega$ and $\mu(A_i) < \infty$ for every i, then μ is called a σ-**finite measure**. If $\mu(\Omega) = 1$, then μ is called a **probability measure**, and in that case, the elements of $\mathcal{A}$ are called **events**. Any probability measure is usually denoted by P.

A set A in $\mathcal{F}$ or $\mathcal{A}$ is called a μ-**null set** if $\mu(A) = 0$. If any relation on Ω holds outside a null set, we say that it holds **almost surely** μ. If the underlying measure is clear from the context, we do not mention it.

Suppose $(\Omega, \mathcal{G}, \mu)$ is a measure space. Let $\mathcal{N}$ be the collection of all subsets of all null sets (of $\mathcal{G}$), and let $\bar{\mathcal{G}}$ be a sub-σ-field of $\mathcal{G}$. Then $\sigma(\bar{\mathcal{G}} \cup \mathcal{N})$ is called the **completion** of $\bar{\mathcal{G}}$. If $\sigma(\mathcal{G} \cup \mathcal{N}) = \mathcal{G}$, then $\mathcal{G}$ is called a **complete σ-field**.

Exercise 1.3.2 Let $\{\mathcal{G}_i\}_{i \geq 1}$ be complete σ-fields. Show that $\bigvee_{i=1}^\infty \mathcal{G}_i$ is also complete.

Definition 1.3.2 (*Trivial σ-field*) Let $(\Omega, \mathcal{G}, P)$ be a probability space. Then a sub-σ-field $\bar{\mathcal{G}}$ is said to be **trivial** if, for any $A \in \bar{\mathcal{G}}$, $P(A)$ is either 0 or 1. ◇

Exercise 1.3.3 Let $(\Omega, \mathcal{A}, \mu)$ be a measure space and $\{A_n\}$ be a sequence of decreasing sets from $\mathcal{G}$ and $\mu(A_n) < \infty$ for some n. Show that $\mu(A_n)$ decreases to $\mu(\bigcap_{i=1}^\infty A_n)$.

Exercise 1.3.4 Let μ be finitely additive on a field $\mathcal{F}$, and $\mu(\Omega) < \infty$. Show that μ is countably additive if, for all sequences $A_n \downarrow \emptyset$ from $\mathcal{F}$, we have $\mu(A_n) \downarrow 0$.

Often a measure is constructed on a σ-field by starting with a measure on a field. The following theorem aids in this construction.

Theorem 1.3.1 (Carathéodory's Extension) *Let μ be a σ-finite measure on $(\Omega, \mathcal{F})$ where $\mathcal{F}$ is a field. Then there is a unique extension of μ to $(\Omega, \sigma(\mathcal{F}))$.* ◆

Definition 1.3.3 For a sequence of sets $\{A_n\}$, define

$$\limsup A_n := \bigcap_{n=1}^\infty \bigcup_{k=n}^\infty A_k, \text{ and } \liminf A_n := \bigcup_{n=1}^\infty \bigcap_{k=n}^\infty A_k.$$

◇

The following lemma is a bread-and-butter tool in probability theory.

Lemma 1.3.1 (Borel–Cantelli) *Let $(\Omega, \mathcal{A}, \mu)$ be a measure space, and $\{A_i\}_{i \geq 1} \subset \mathcal{A}$. Then $\sum_{n=1}^\infty \mu(A_n) < \infty$ implies $\mu(\limsup A_n) = 0$.* ◆

1.4 Measurable Function and Integration

Let $(\Omega, \mathcal{G})$ be a measurable space. For $A \in \mathcal{A}$, its **indicator function**, $\mathbf{1}_A$, is 1 and 0 on A and A^c, respectively. Suppose $f : \Omega \to M$ is a function where M is a topological space equipped with the Borel σ-field $\mathcal{B}$. If for some finite k, f is of the form $f := \sum_{i=1}^k m_i \mathbf{1}_{A_i}$ where $\{m_i\}$ are from M and $\{A_i\}$ are from $\mathcal{A}$ then f is called a **simple function**.

In general, $f : \Omega \to M$ is said to be **Borel measurable**, or just **measurable**, if $f^{-1}(B) \in \mathcal{A}$ for every $B \in \mathcal{B}$. All our functions will be measurable, even if we do not mention it explicitly.

Real-valued Borel measurable functions on a probability space are called **random variables**, and are usually denoted by X, Y, X_i, Y_n, etc.

1.4 Measurable Function and Integration

A function f is **extended real-valued**, if it takes values in $\bar{\mathbb{R}}$. We may not explicitly mention whether a function is extended real valued or not, when it is clear from the context.

For any such f, the functions f^+ and f^- will denote its positive and negative parts, so that $f = f^+ - f^-$ and $|f| = f^+ + f^-$. All these functions are also measurable.

Exercise 1.4.1 Show that the Borel σ-field of $\mathbb{R}^{\mathbb{N}}$ is the smallest σ-field on $\mathbb{R}^{\mathbb{N}}$ with respect to which, for each $n \in \mathbb{N}$, the coordinate function $\theta_n : \mathbb{R}^{\mathbb{N}} \to \mathbb{R}$, defined by $\theta_n(x) := x_n$ (where $x = (x_1, \ldots, x_n, \ldots)$), is real-valued Borel measurable.

Suppose $(\Omega, \mathcal{G}, \mu)$ is a measure space. Then for any non-negative measurable function f, its integral $\int_\Omega f d\mu$ (we also write $\int_\Omega f(\omega)\mu(d\omega)$) with respect to μ is a **non-negative linear functional** which satisfies all the natural properties of an integral and $\int_\Omega 1 d\mu = \mu(\Omega)$.

For any extended real-valued measurable function f, $\int_\Omega f d\mu$ is said to **exist**, if at least one of the integrals $\int_\Omega f^+ d\mu$ or $\int_\Omega f^- d\mu$ is finite. If both integrals are finite, then f is said to be **integrable**. For any $A \in \mathcal{A}$, $\int_A f d\mu$ is nothing but $\int_\Omega f 1_A d\mu$. If $(\Omega, \mathcal{A}, P)$ is a probability space and X is integrable, then we write $\int_\Omega X dP$ as $E(X)$, and call it the **expectation** or the **mean** of X.

Definition 1.4.1 (*Product space*) Let $(\Omega_i, \mathcal{A}_i)_{1 \leq i \leq k}$ be measurable spaces. The **product measurable space** is defined as $(\Omega, \mathcal{A})$, where $\Omega := \Omega_1 \times \cdots \times \Omega_k$, and $\mathcal{A}$ is the smallest σ-field containing the rectangles $A_1 \times \cdots \times A_k$, $A_i \in \mathcal{A}_i$, $1 \leq i \leq k$. The σ-field $\mathcal{A}$ is also written as $\mathcal{A} =: \mathcal{A}_1 \otimes \cdots \otimes \mathcal{A}_k$.

A function $\mu_2 : \Omega_1 \times \mathcal{A}_2 \to [0, \infty]$ is said to be a **transition measure** if for every $\omega_1 \in \Omega_1$, $\mu_2(\omega_1, \cdot)$ is a measure on $(\Omega_2, \mathcal{A}_2)$, and for every $A_2 \in \mathcal{A}_2$, $\mu_2(\cdot, A_2)$ is a Borel measurable function on $(\Omega_1, \mathcal{A}_1)$. This transition measure is said to be **uniformly σ-finite**, if there exists a partition $\{A_i\}$ of Ω_2 of sets from $\mathcal{A}_2$ such that $\sup_{\omega_1 \in \Omega_1} \mu_2(\omega_1, A_i) < \infty$ for all $i \geq 1$ and all $A_2 \in \mathcal{A}_2$. $\diamond$

The following result on integrals will be used repeatedly.

Theorem 1.4.1 (Fubini's theorem) *Let $(\Omega_i, \mathcal{A}_i)_{1 \leq i \leq 2}$ be measurable spaces. Suppose μ_1 is a σ-finite measure on $(\Omega_1, \mathcal{A}_1)$, and μ_2 is a uniformly σ-finite transition measure on $\Omega_1 \times \mathcal{A}_2$. Let $\Omega := \Omega_1 \times \Omega_2$ and $\mathcal{A} := \mathcal{A}_1 \otimes \mathcal{A}_2$. Let $f : \Omega \to \bar{\mathbb{R}}$ be a Borel measurable function. Then the following hold:*

(i) If f is non-negative, then

$$g(\omega_1) := \int_{\Omega_2} f(\omega_1, \omega_2)\mu_2(\omega_1, d\omega_2), \quad \omega_1 \in \Omega_1, \tag{1.1}$$

exists, and defines a Borel measurable function on Ω_1. Further,

$$\int_\Omega f d\mu = \int_{\Omega_1} g(\omega_1)\mu_1(d\omega_1)$$
$$= \int_{\Omega_1} \left[\int_{\Omega_2} f(\omega_1, \omega_2)\mu_2(\omega_1, d\omega_2)\right]\mu_1(d\omega_1). \tag{1.2}$$

(ii) If $\int_\Omega f d\mu$ exists (respectively, finite), then $g(\omega_1)$ defined in (1.1) does exist (respectively, finite) almost surely μ_1, and defines a Borel measurable function, if it is defined to be, say, 0 on the relevant null set. Moreover (1.2) holds. ◆

1.5 Convergence

We first define two notions of convergence of measurable functions.

Definition 1.5.1 Let $(\Omega, \mathcal{G}, \mu)$ be a measure space and $\{f_n\}$ and f be Borel measurable.

(a) The sequence $\{f_n\}$ is said to **converge almost surely** to f, if there is a μ-null set N, such that $f_n(\omega) \to f(\omega)$ for all $\omega \in N^c$. We say f_n converges to f almost surely μ.

(b) The sequence $\{f_n\}$ is said to **converge in measure** to f (we write $f_n \xrightarrow{\mu} f$), if for every $\varepsilon > 0$, $\lim_{n\to\infty} \mu\{|f_n - f| > \varepsilon\} = 0$. Convergence in probability is the same as convergence in measure when the underlying measure is a probability measure. ◇

The most useful convergence results on integrals are listed below. These will be used repeatedly later.

Theorem 1.5.1 (Convergence of integrals) *Suppose $\{g_n\}_{n\geq 0}$ are measurable functions on $(\Omega, \mathcal{G}, \mu)$.*

*(i) **Fatou's lemma**. If $\{g_n\}$ are non-negative, then*

$$\int_\Omega \liminf_{n\to\infty} g_n d\mu \leq \liminf_{n\to\infty} \int_\Omega g_n d\mu.$$

*(ii) **Monotone convergence theorem** (MCT). If $g_n \uparrow g$ almost surely, and $\{g_n\}$ are non-negative for all n, then $\int_\Omega g_n d\mu \uparrow \int_\Omega g d\mu$.*

*(iii) **Dominated convergence theorem** (DCT). Suppose $g_n \to g$ a.s., $|g_n| \leq f$ for all $n \geq 0$ and $\int_\Omega f d\mu < \infty$. Then $\int_\Omega g_n d\mu \to \int_\Omega g d\mu$.*

*(iv) **DCT in probability**. Suppose $(\Omega, \mathcal{G}, P)$ is a probability space and X_n, X, Y are random variables defined on Ω. If $X_n \xrightarrow{P} X$, where $|X_n| \leq Y$ for all $n \geq 0$ and Y is integrable, then $E(X_n) \to E(X)$.* ◆

1.6 L^p Space

If two random variables X and Y are equal almost surely, then we treat them as being equal. Then we have the following definition.

1.6 L^p Space

Definition 1.6.1 (L^p *space*, $1 \leq p < \infty$) Suppose $(\Omega, \mathcal{G}, P)$ is a probability space. Then the $L^p(P)$-space (L^p-space in short) is defined as

$$L^p(\Omega, \mathcal{G}, P) := \{X : E(|X|^p) < \infty\}.$$

Then $d_p(X, Y) := \left[E(|X - Y|^p)\right]^{1/p}$ is a **complete metric** on L^p. ◇

If X, Y are in L^2, then their **covariance** is defined as

$$\mathrm{Cov}(X, Y) := E\big((X - E(X))(Y - E(Y))\big) = E(XY) - [E(X)][E(Y)].$$

For $X \in L^2$, $\mathrm{Cov}(X, X)$ is called the **variance** of X, written as $\mathrm{Var}(X)$.

The following inequalities are well known.

Theorem 1.6.1 *(i)* (**Markov inequality**) *Let X be a real-valued random variable. Then for any $\varepsilon > 0$,*

$$P\{|X| > \varepsilon\} \leq \frac{E(|X|)}{\varepsilon}.$$

(ii) (**Chebyshev's inequality**). *For any real-valued random variable X with finite mean μ, and any $\varepsilon > 0$,*

$$P\{|X - \mu| > \varepsilon\} \leq \frac{\mathrm{Var}(X)}{\varepsilon^2}.$$

(iii) (**Cauchy–Schwarz inequality**). *Suppose X, Y are in L^2. Then*

$$|E(XY)|^2 \leq E(X^2)E(Y^2).$$

(iv) (**Hölder's inequality**). *Suppose $1 < p, q < \infty$ with $\frac{1}{p} + \frac{1}{q} = 1$. Suppose $X \in L^p$ and $Y \in L^q$. Then $XY \in L^1$, and*

$$|E(XY)| \leq \left[E(|X|^p)\right]^{1/p} \left[E(|Y|^q)\right]^{1/q}.$$

(v) (**Lyapunov's inequality**). *If $0 < p \leq q < \infty$, then for any random variable X,*

$$\left[E\left(|X|^p\right)\right]^{1/p} \leq \left[E\left(|X|^q\right)\right]^{1/q}.$$

♦

Exercise 1.6.1 (a) Suppose $d_p(X_n, X) \to 0$ for some $p \geq 1$. Then show that $X_n \xrightarrow{P} X$.

(b) Show that convergence in L^p implies convergence in L^q if $1 \leq q < p$.

1.7 Distribution Function

Definition 1.7.1 (*Distribution function*) Any function $F : \mathbb{R} \to \mathbb{R}$ is called a **distribution function** if

(a) F is non-decreasing: $F(a) \leq F(b)$ for all $a \leq b$, $a, b \in \mathbb{R}$, and

(b) F is right continuous: $\lim_{y \downarrow x} F(y) = F(x)$ for every $x \in \mathbb{R}$. ◇

Remark 1.7.1 (i) A distribution function can take negative values, and can be unbounded. For example, $F(x) = x$, $x \in \mathbb{R}$ is a distribution function.

(ii) If F is a distribution function, then $F(x-) = \lim_{y \to x,\ y < x} F(y)$ exists for each $x \in \mathbb{R}$.

(iii) We shall soon see that, given any distribution function, there is an associated measure on $\mathcal{B}(\mathbb{R})$ with certain properties and vice versa. ●

Exercise 1.7.1 Let f be a non-negative Riemann integrable function on $\mathbb{R}$. Show that F defined below is a distribution function.

$$F(x) := \int_{-\infty}^{x} f(y) dy, \ x \in \mathbb{R}.$$

1.8 Lebesgue–Stieltjes Measure

Carathéodory's extension theorem will lead us to the Lebesgue measure (the length measure) on $\mathbb{R}$. Indeed we shall exhibit a more general class of measures on $\mathbb{R}$, which are called Lebesgue–Stieltjes measures, and these are intimately tied to distribution functions.

Definition 1.8.1 (*Lebesgue–Stieltjes measure*) A measure μ on $\mathcal{B}(\mathbb{R})$ is called **Lebesgue–Stieltjes**, if $\mu(I) < \infty$ for every bounded interval I of $\mathbb{R}$. ◇

Note that any Lebesgue–Stieltjes measure is automatically σ-finite.

Theorem 1.8.1 (From a Lebesgue–Stieltjes measure to a distribution function) *Suppose μ is a Lebesgue–Stieltjes measure on $\mathcal{B}(\mathbb{R})$.*

(i) Fix $F(0)$ arbitrarily. Then F defined below is a distribution function.

$$F(x) := \begin{cases} F(0) + \mu(0, x] & \text{if } x > 0, \\ F(0) - \mu(x, 0] & \text{if } x < 0. \end{cases} \quad (1.3)$$

(ii) If μ is a finite measure, then we may choose $F(0) = \mu(-\infty, 0]$. With this choice (1.3) reduces to

$$F(x) = \mu(-\infty, x], \text{ for all } x \in \mathbb{R}. \quad (1.4)$$

◆

1.8 Lebesgue-Stieltjes Measure

Proof of Theorem 1.8.1 Suppose $a < b$. Then from (1.3),

$$F(b) - F(a) = \mu(a, b] \geq 0,$$

establishing that F is non-decreasing.

Fix $x \in \mathbb{R}$ and $x_n \downarrow x$. Since μ is finite on every sub-interval, using Exercise 1.3.3 on continuity from above,

$$F(x_n) - F(x) = \mu(x, x_n] \downarrow \mu(\emptyset) = 0,$$

establishing right continuity. □

Remark 1.8.1 Whenever μ is a finite measure, we shall always choose the corresponding distribution function as described in (1.4). ●

Theorem 1.8.2 (From distribution function to Lebesgue–Stieltjes measure) *Suppose F is a distribution function on $\mathbb{R}$. Define μ on right closed intervals as, $\mu(a, b] := F(b) - F(a)$, $a < b$. Then μ has a unique extension to $\mathcal{B}(\mathbb{R})$, as a Lebesgue–Stieltjes measure.* ◆

Proof The idea is to first define a finitely additive set function on a suitable semi-field, extend it to a field, and then show that on this field it is countably additive. Then we can use Carathéodory's extension Theorem 1.3.1.

First, F is extended to the function $\bar{F} : \bar{\mathbb{R}} \to \bar{\mathbb{R}}$ by defining

$$F(\infty) = \lim_{x \to \infty} F(x), \quad F(-\infty) = \lim_{x \to -\infty} F(x).$$

Note that both the above limits exist but they may equal ∞ and $-\infty$, respectively. Define

$$\bar{S} := \{(a, b] \cap \mathbb{R} : a, b \in \bar{\mathbb{R}}, a \leq b\},$$
$$\bar{\mathcal{F}} := \{\cup_{k=1}^{n} I_k : I_k \in \bar{S} \text{ for } k = 1, \ldots, n \text{ and are disjoint}\}.$$

From Exercise 1.2.1, $\bar{S}$ is a semi-field, and $\bar{\mathcal{F}}$ is a field of subsets of $\mathbb{R}$. Note that $\bar{\mathcal{F}}$ is a superset of the smallest field containing all bounded intervals of the form $(a, b]$ of $\mathbb{R}$.

Now define μ on $\bar{S}$ by

$$\mu((a, b] \cap \mathbb{R}) := F(b) - F(a), \ a, b \in \bar{\mathbb{R}}, \ a < b, \qquad (1.5)$$
$$\mu(\emptyset) := 0.$$

An immediate observation is that

$$\mu\left(\bigcup_{j=1}^{n} I_j\right) = \sum_{j=1}^{n} \mu(I_j), \ I_1, \ldots, I_n \in \bar{S} \text{ are disjoint, and } \bigcup_{j=1}^{n} I_j \in \bar{S}. \qquad (1.6)$$

Extend μ to $\bar{\mathcal{F}}$ by

$$\mu(\bigcup_{k=1}^{n} I_k) := \sum_{k=1}^{n} \mu(I_k), \quad I_1, \ldots, I_n \in \bar{\mathcal{S}}, \text{ and they are disjoint.} \quad (1.7)$$

Note that $\cup_{k=1}^n I_k$ has alternate descriptions as $\cup_{k=1}^t J_k$ where J_k are disjoint elements of $\bar{\mathcal{S}}$. The definition (1.7) is consistent because (1.6) shows in this case

$$\sum_{k=1}^{n} \mu(I_k) = \sum_{j=1}^{n} \sum_{k=1}^{t} \mu(I_j \cap J_k) = \sum_{k=1}^{t} \mu(J_k).$$

Further, (1.6), (1.7) and the above argument also show μ is finitely additive on $\bar{\mathcal{F}}$, that is,

$$\mu(\bigcup_{i=1}^{n} A_i) = \sum_{i=1}^{n} \mu(A_i) \quad \text{for any disjoint } A_1, \ldots, A_n \in \bar{\mathcal{F}}. \quad (1.8)$$

Easy consequences of the above are the following:

$$\mu(A) \leq \mu(B), \quad A, B \in \bar{\mathcal{F}}, \ A \subset B, \quad (1.9)$$

$$\mu(A_1 \cup \cdots \cup A_n) \leq \sum_{i=1}^{n} \mu(A_i), \quad A_1, \ldots, A_n \in \bar{\mathcal{F}}. \quad (1.10)$$

Toward showing μ is countably additive on $\bar{\mathcal{S}}$, let $I_1, I_2, I_3, \ldots \in \bar{\mathcal{S}}$ be disjoint and $I = \cup_{i=1}^{\infty} \in \bar{\mathcal{S}}$.

For $n \geq 1$, (1.9) shows that

$$\mu(I) \geq \mu(I_1 \cup \cdots \cup I_n) = \sum_{j=1}^{n} \mu(I_j),$$

where the equality above follows from (1.8). Thus,

$$\sum_{j=1}^{\infty} \mu(I_j) \leq \mu(I). \quad (1.11)$$

For the reverse inequality, write $I = (a, b] \cap \mathbb{R}$ for some $a, b \in \bar{\mathbb{R}}$ with $a < b$. Similarly for $j - 1, 2, \ldots$, let $I_j = (a_j, b_j] \cap \mathbb{R}$ with $\infty \leq a_j < b_j \leq \infty$.

Fix $\delta > 0$ and choose $\varepsilon_j > 0$ so that

$$F(b_j + \varepsilon_j) \leq F(b_j) + 2^{-j}\delta, \quad j = 1, 2, \ldots;$$

1.8 Lebesgue-Stieltjes Measure

such a choice is trivial if $b_j = \infty$, otherwise it is due to right continuity of F.

Observe that for all $j = 1, 2, \ldots, I_j \subset (a_j, b_j + \varepsilon_j)$. Thus, for $a < a' < b' \leq b$ with $b' < \infty$, $[a', b'] \subset \cup_{j=1}^{\infty}(a_j, b_j + \varepsilon_j)$.

As $[a', b']$ is compact, using Heine–Borel theorem, for some n, $[a', b'] \subset \cup_{j=1}^{n}(a_j, b_j + \varepsilon_j)$.

Therefore,

$$F(b') - F(a') = \mu(a', b'] \leq \mu\left(\cup_{j=1}^{n}\left((a_j, b_j + \varepsilon_j] \cap \mathbb{R}\right)\right) \text{ (by (1.9))}$$

$$\leq \sum_{j=1}^{n}\left(F(b_j + \varepsilon_j) - F(a_j)\right) \text{ (by (1.10))}$$

$$\leq \sum_{j=1}^{\infty}\left(2^{-j}\delta + F(b_j) - F(a_j)\right) \text{ (choice of } \varepsilon_j)$$

$$= \delta + \sum_{j=1}^{\infty}\mu(I_j).$$

As δ is arbitrary, we get

$$F(b') - F(a') \leq \sum_{j=1}^{\infty}\mu(I_j). \tag{1.12}$$

As $a' \downarrow a$, $F(a') \to F(a)$, which is again a consequence of right continuity of F when $a > -\infty$, and otherwise follows from the definition of $F(-\infty)$. Thus the inequality (1.12) is preserved if a' is replaced by a. In the case $b < \infty$, put $b' = b$, otherwise let $b' \to \infty$ to get

$$F(b) - F(a) \leq \sum_{j=1}^{\infty}\mu(I_j).$$

This along with (1.11) shows μ is countably additive on $\bar{\mathcal{S}}$. Its countable additivity on $\bar{\mathcal{F}}$ is left as an exercise. Evidently, μ is σ-finite on $\bar{\mathcal{F}}$ because $\mu(n, n+1] < \infty$ for all $n \in \mathbb{Z}$. Recalling that $\bar{\mathcal{F}}$ is a field of subsets of $\mathbb{R}$, Carathéodory's extension Theorem 1.3.1 shows that μ can be extended to a unique measure on $\mathcal{B}(\mathbb{R})$. This completes the proof. □

1.9 Lebesgue Measure

We can now define the "length measure" on $\mathbb{R}$. In Theorem 1.8.2, let $F(x) = x + c$, where c is any constant. Then the corresponding measure is uniquely defined, irrespective of the choice of c. This is called the Lebesgue measure.

Definition 1.9.1 (*Lebesgue measure on* $\mathbb{R}$) The measure λ such that $\lambda(a, b] = b - a$ for all $a < b \in \mathbb{R}$ is said to be the **Lebesgue measure** on $\mathcal{B}(\mathbb{R})$. Usually we simply say λ is the Lebesgue measure on $\mathbb{R}$. ◇

Integral of a function f with respect to the Lebesgue measure is usually written as $\int f d\lambda$, $\int f(x)\lambda(dx)$ or $\int f(x)dx$.

The above results on Lebesgue–Stieltjes measures and distribution functions on $\mathbb{R}$ can be easily extended to those on $\mathbb{R}^n$. However, to define distribution functions on $\mathbb{R}^n$, the notion of non-decreasing has to be defined appropriately. The details will not be presented in this book (see Billingsley (1995) for further details).

Here is an important special case. Let F on $\mathbb{R}^n$ be of the form

$$F(a_1, a_2, \ldots, a_n) = F_1(a_1) \cdots F_n(a_n), \ a_i \in \mathbb{R}, \ i = 1, 2, \ldots, n,$$

where each $F_i(\cdot)$ is a distribution function. Then the construction of the measure is a bit simpler. One can start with rectangles that are product of right-semi-closed intervals, and define μ in the natural way. The existence of Lebesgue measure on $\mathbb{R}^n$ then follows by choosing $F_i(x) = x$, for each $1 \leq i \leq n$.

This is also really a special case of the **product measure construction**, starting with the Lebesgue measure on $\mathbb{R}$. We shall not cover details of product measures in this book.

Definition 1.9.2 (*Lebesgue measure on* $\mathbb{R}^n$) The Lebesgue measure λ_n on $\mathcal{B}(\mathbb{R}^n)$ is the unique measure such that

$$\lambda_n\big((a_1, b_1] \times \cdots \times (a_n, b_n]\big) = \prod_{i=1}^{n}(b_i - a_i),$$

for all $a_i < b_i$, $a_i, b_i \in \mathbb{R}$, $1 \leq i \leq n$. ◇

Let μ be a measure on $\mathcal{B}(\mathbb{R}^n)$, and B be a Borel set. It is of interest to understand when $\mu(B)$ can be approximated from above or from below by compact or open sets.

Theorem 1.9.1 (Approximation) *Let μ be a measure on $\mathcal{B}(\mathbb{R}^n)$.*
(i) (Approximation from within) *If μ is σ-finite, then*

$$\mu(B) = \sup\{\mu(K) : K \subseteq B, \ K \text{ compact}\}, \text{ for every } B \in \mathcal{B}(\mathbb{R}^n). \quad (1.13)$$

(ii) (Approximation from above) *If μ is finite, then*

1.9 Lebesgue Measure

$$\mu(B) = \inf\{\mu(V) : V \supseteq B, \ V \text{ open}\}, \text{ for every } B \in \mathcal{B}(\mathbb{R}^n). \tag{1.14}$$

(iii) If μ is a Lebesgue–Stieltjes measure, then (1.14) still holds. ◆

Proof (i) First suppose μ is finite. Let

$$\mathcal{M} := \{B \in \mathcal{B}(\mathbb{R}^n) : (1.13) \text{ holds}\}. \tag{1.15}$$

Then we claim that $\mathcal{M}$ is a monotone class. First let $B_n \in \mathcal{M}$, $B_n \uparrow B$. Fix $\varepsilon > 0$. Let $D_n \subseteq B_n$ be compact sets such that $\mu(B_n) \leq \mu(D_n) + \varepsilon$. Let $K_n := \cup_{k=1}^n D_k$. Then $\{K_n\}$ is a non-decreasing sequence of compact sets. Moreover,

$$K_n \subseteq B_n \text{ and } (B_n \setminus K_n) \subseteq (B_n \setminus D_n) \text{ for all } n \geq 1.$$

Hence $\mu(B_n \setminus K_n) \leq \mu(B_n \setminus D_n) \leq \varepsilon$. This implies

$$\lim_{n \to \infty} \mu(K_n) \leq \mu(B) = \lim_{n \to \infty} \mu(B_n) \leq \lim_{n \to \infty} \mu(K_n) + \varepsilon,$$

so that (1.13) holds for B, and hence $B \in \mathcal{M}$. Now suppose $B_n \in \mathcal{M}$, $B_n \downarrow B$. Let $K_n \subset B_n$ compact be such that

$$\mu(B_n) \leq \mu(K_n) + \frac{\varepsilon}{2^n}.$$

Let $K := \cap_{n=1}^\infty K_n$. Note that K is compact. Then

$$\mu(B) - \mu(K) = \mu(B \setminus K) \text{ (since } K \subseteq B\text{)}$$
$$\leq \mu\bigl(\cup_{n=1}^\infty (B_n \setminus K_n)\bigr) \text{ (since } B \subseteq \cup_{n=1}^\infty B_n\text{)}$$
$$\leq \sum_{n=1}^\infty \mu(B_n \setminus K_n) \leq \varepsilon \text{ (countable sub-additivity)}.$$

Hence $B \in \mathcal{M}$. Thus $\mathcal{M}$ is a monotone class. Note that $\mathcal{M}$ contains all right-semi-closed intervals, and their finite disjoint unions. But then this collection forms a field that generates $\mathcal{B}(\mathbb{R}^n)$. By the Monotone Class Theorem 1.1.1, $\mathcal{M} = \mathcal{B}(\mathbb{R}^n)$. This proves Part (i) when μ is finite.

Now let μ be σ-finite. Get $B_n \in \mathcal{B}(\mathbb{R}^n)$ such that $\mu(B_n) < \infty$ and $B_n \uparrow B$. Each B_n can be approximated from within by compact sets. To see this, consider the finite measures $\mu_n(A) = \mu(A \cap B_n)$, $A \in \mathcal{B}(\mathbb{R}^n)$ and proceed as in the finite case (we need to observe that finite unions of compact sets are compact). This completely proves Part (i).

(ii) We have

$$\mu(B) \leq \inf\{\mu(V) : V \supseteq B, \ V \text{ open}\}$$
$$\leq \inf\{\mu(W) : W \supseteq B, \ W = K^c, \ K \text{ compact}\}$$

$$\begin{aligned}
&= \inf\left\{\mu(\mathbb{R}) - \mu(W^c) : W^c \subseteq B^c,\ W^c \text{ compact}\right\} \\
&= \mu(\mathbb{R}) - \sup\left\{\mu(K) : K \subseteq B^c,\ K \text{ compact}\right\} \\
&= \mu(\mathbb{R}) - \mu(B^c) \text{ (by (i), since } \mu \text{ is finite)} \\
&= \mu(B).
\end{aligned}$$

(iii) Let $\mathbb{R}^n = \cup_{n=1}^{\infty} B_n$ where $\{B_n\}$ are disjoint and bounded. Then for each n, $B_n \subseteq C_n$ for some bounded *open* set C_n. Define finite measures

$$\mu_k(A) := \mu(A \cap C_k),\ A \in \mathcal{B}(\mathbb{R}^n).$$

Fix $\varepsilon > 0$. If B is a Borel subset of B_k, then by (ii), there is an open set $W_k \supseteq B$ such that

$$\mu_k(W_k) \leq \mu_k(B) + \frac{\varepsilon}{2^k}. \tag{1.16}$$

Note that $V_k = W_k \cap C_k$ is an open set. Further, $B \cap C_k = B$ since $B \subseteq B_k \subseteq C_k$. Hence

$$\begin{aligned}
\mu(V_k) &= \mu_k(W_k) \text{ (by definition of } \mu_k) \\
&\leq \mu_k(B) + \frac{\varepsilon}{2^k} \text{ (by (1.16))} \\
&= \mu(B) + \frac{\varepsilon}{2^k} \text{ (since } B \subseteq C_k).
\end{aligned}$$

Now fix any $A \in \mathcal{B}(\mathbb{R}^n)$. By the conclusion reached above, let V_k be an open set such that

$$V_k \supseteq A \cap B_k \text{ and } \mu(V_k) \leq \mu(A \cap B_k) + \frac{\varepsilon}{2^k} \text{ for all } k.$$

Let $V := \cup_{n=1}^{\infty} V_n$. Then V is open, $V \supseteq A$. The above inequality yields

$$\mu(V) \leq \sum_{n=1}^{\infty} \mu(V_n) \leq \mu(A) + \varepsilon.$$

This proves Part (iii), and the proof of the theorem is complete. □

1.10 Probability Distribution

Definition 1.10.1 Any $\mathbb{R}^n$-valued random variable on $(\Omega, \mathcal{A}, P)$ induces a **probability measure** P_X on $(\mathbb{R}^n, \mathcal{B}(\mathbb{R}^n))$ by the following formula:

$$P_X(B) := P\{\omega \in \Omega : X(\omega) \in B\}.$$

Often we write $P(X \in \cdot)$ for P_X. This measure on $\mathcal{B}(\mathbb{R}^n)$ is called the **probability distribution of** X. If there exists a non-negative Lebesgue integrable function f on $\mathbb{R}^n$ such that

1.10 Probability Distribution

$$P_X(B) = \int_B f(x)\lambda(dx) \text{ for all } B \in \mathcal{B}(\mathbb{R}^n),$$

then f is called the **probability density of** X. If X and Y (possibly defined on different probability spaces) have the same probability distribution, then we write it as $X \stackrel{d}{=} Y$. ◇

Exercise 1.10.1 Let X be a random variable on $(\Omega, \mathcal{G}, P)$ and $g : \mathbb{R} \to \mathbb{R}$ be Borel measurable. Show that $g(X) : \Omega \to \mathbb{R}$ is measurable. Further,

$$E(g(X)) = \int_\Omega g(X(\omega))P(d\omega) = \int_\mathbb{R} g(x)P_X(dx).$$

That is, if one of the above integrals exists, then the other also exists, and they are equal. Thus $E(g(X))$ can be calculated by using either of the two probability measures, P or P_X.

Exercise 1.10.2 Let X be a random variable. Using Fubini's theorem, show that

$$E(|X|) = \int_0^\infty P\{|X| > x\}\lambda(dx) = \int_0^\infty P\{|X| \geq x\}\lambda(dx).$$

Example 1.10.1 (a) A random variable Y is said to have the **Poisson distribution** with parameter $0 < \lambda < \infty$, if

$$P\{Y = k\} = e^{-\lambda}\frac{\lambda^k}{k!}, \ k = 0, 1, \ldots.$$

(b) A random variable X is said to have the **Binomial distribution** with parameters (n, p), where n is an integer, and $0 \leq p \leq 1$, if

$$P\{X = k\} = \binom{n}{k}p^k(1-p)^{n-k}, \ k = 0, 1, \ldots, n.$$

In the particular case where $n = 1$, X is said to have a **Bernoulli distribution** with success probability p.

(c) A random variable X is said to have the **Gaussian distribution** with mean $\mu \in \mathbb{R}$, and variance $\sigma^2 > 0$ if

$$P\{X \in B\} = \frac{1}{\sqrt{2\pi}\sigma}\int_B e^{-(x-\mu)^2/(2\sigma^2)}\lambda(dx), \text{ for all } B \in \mathcal{B}(\mathbb{R}).$$

This is a probability distribution. Further $E(X) = \mu$ and $\text{Var}(X) = \sigma^2$. For the case $\sigma^2 = 0$, $\mathbb{P}\{X = \mu\} = 1$, and we continue to call X Gaussian. We denote the Gaussian distribution or its CDF as $N(\mu, \sigma^2)$ and say $X \sim N(\mu, \sigma^2)$.

(d) A random vector X is **multivariate Gaussian** with expectation vector μ and covariance matrix Σ if, for every vector of constants c, $c'X$ has a Gaussian distribution with mean $c'\mu$ and variance $c'\Sigma c$.

(e) X is said to be **uniformly distributed** on the interval $[0, 1]$, if its probability measure is the Lebesgue measure on $[0, 1]$. We write $X \sim U[0, 1]$. The probability density function of X is the constant function 1 on the interval $[0, 1]$. ▲

Definition 1.10.2 A distribution function F is a **cumulative distribution function** (CDF) if $\lim_{x \to -\infty} F(x) = 0$, and $\lim_{x \to \infty} F(x) = 1$. ◇

Exercise 1.10.3 Suppose for $\alpha, p > 0$,

$$F(x) := \begin{cases} \int_0^x \frac{\alpha^p}{\Gamma(p)} e^{-\alpha y} y^{p-1} \lambda(dy) & \text{if } x > 0, \\ 0 & \text{otherwise}. \end{cases} \quad (1.17)$$

Show that F is a CDF. This is known as the **gamma distribution**. The special case when $p = 1$ is known as the **exponential distribution**.

Exercise 1.10.4 Let P be the measure with $P\{i\} = 2^{-i}, i \geq 1$. Define $F(x) := \sum_{i \leq x} P\{i\}$. Show that F is a CDF and identify its nature.

Exercise 1.10.5 Suppose $(\Omega, \mathcal{G}, P)$ is a probability space, and X is a real-valued random variable. Define

$$F_X(x) := P\{X^{-1}(-\infty, x]\}, x \in \mathbb{R}.$$

Show that $F_X(\cdot)$ is a CDF, known as the **CDF of** X.

Definition 1.10.3 A sequence $\{X_n\}$ of real-valued random variables on $\{(\Omega_n, \mathcal{A}_n, P_n)\}$ with CDFs $\{F_n\}$ is said to **converge in distribution** if there is a CDF F such that $F_n(x) \to F(x)$ at every continuity point x of F. If Z has CDF F, then we write this convergence as $X_n \Rightarrow Z$, or as $F_n \Rightarrow F$. We also say X_n (or F_n) **converge weakly** to X (or F). ◇

Exercise 1.10.6 Show that convergence in probability implies convergence in distribution.

Exercise 1.10.7 Suppose $\{X_n\}$ converges in distribution to X, and all of the variables have finite means. Show by an example that, $E(X_n)$ need not converge to $E(X)$.

Exercise 1.10.8 Let $\{X_n\}$ be a sequence of Binomial random variables with parameters (n, p_n) where $np_n \to \lambda$. Show that $\{X_n\}$ converges in distribution, and identify the limit.

Exercise 1.10.9 Let $\{X_n\}$ and $\{Y_n\}$ be two sequences of random variables on a probability space $(\Omega, \mathcal{G}, P)$. Suppose $X_n \Rightarrow X$, and $Y_n \to 0$ in probability. Show that $X_n + Y_n \Rightarrow X$.

Exercise 1.10.10 Let $\{X_n\}$ and $\{Y_n\}$ be sequences of random variables on $(\Omega, \mathcal{G}, P)$. Suppose that $Y_n \to 0$ in probability and $\{X_n\}$ is bounded by K. Show that $X_n Y_n$ converges to 0 in probability.

Exercise 1.10.11 Let $\{X_n\}$ and $\{Y_n\}$ be sequences of random variables on $(\Omega, \mathcal{G}, P)$ and let $\{A_n\}$ be a sequence of measurable sets. Suppose $X_n \Rightarrow X$, $\{Y_n\}$ is bounded and $\mathbf{1}_{A_n} \to 1$ in probability. Show that $Y_n \mathbf{1}_{A_n^c} + X_n \mathbf{1}_{A_n} \Rightarrow X$.

1.11 Characteristic Function

Definition 1.11.1 Let X be a real-valued r.v. with CDF F. Then the **characteristic function** of X or of F is defined as (with $\iota = \sqrt{-1}$),

$$\phi_X(t) = \phi_F(t) := \int_{\mathbb{R}} e^{\iota t x} F(dx) = E(e^{\iota t X}), \ t \in \mathbb{R}.$$

◇

Exercise 1.11.1 Verify that the characteristic function is indeed well defined for all $t \in \mathbb{R}$. Moreover, $\phi_X(0) = 1$, $|\phi_X(t)| \leq 1$ for all t, and $\phi_X(\cdot)$ is uniformly continuous on $\mathbb{R}$.

The classic books of Lukacs (1970) and Ramachandran (1967) contain extensive results on characteristic functions. The concept of characteristic function is extremely useful in the study of CDFs due to the following theorem of Lévy, which connects convergence in distribution with characteristic function. We shall use this result in Chap. 7.

Theorem 1.11.1 (Lévy continuity theorem) *Suppose $\{X_n\}$ is a sequence of real-valued random variables. Let X be another random variable. Then $\phi_{X_n}(t) \to \phi_X(t)$ for all $t \in \mathbb{R}$ if and only if, $\{X_n\}$ converges in distribution to X.* ◆

Exercise 1.11.2 Show that if $Z \sim N(\mu, \sigma^2)$, then its characteristic function is given by $\phi_Z(t) = e^{it\mu - t^2\sigma^2/2}$. Find the characteristic functions of the Binomial and the Poisson random variables.

Exercise 1.11.3 Solve Exercise 1.10.8 using Theorem 1.11.1.

Exercise 1.11.4 Suppose that a sequence of Gaussian random variables converges in distribution. Show that then the limit distribution must also be Gaussian.

Exercise 1.11.5 Let $\{X_n\}$ be a sequence of Binomial random variables with parameters p and n. Show that $(X_n - np)/\sqrt{np(1-p)} \Rightarrow Z$ where $Z \sim N(0, 1)$.

1.12 Independence

Recall that arbitrary intersection of σ-fields is a σ-field. Let $\mathcal{X}$ be any collection of random variables defined on $(\Omega, \mathcal{G}, P)$. Then $\sigma(\mathcal{X})$ will denote the smallest sub-σ-field of $\mathcal{A}$ with respect to which all these variables are measurable. If a real-valued random variable Y is measurable with respect to $\sigma(\mathcal{X})$, we write $Y \in \sigma(\mathcal{X})$ by an abuse of notation.

Exercise 1.12.1 Let X be a random variable on $(\Omega, \mathcal{A}, P)$ which assumes only countably many values $\{x_i\}$. Then X is said to be a **discrete** random variable. Describe $\sigma(X)$.

Exercise 1.12.2 Let X be an $\mathbb{R}^n$-valued random variable on $(\Omega, \mathcal{A}, P)$. Show that a random variable Y defined on Ω is measurable with respect to $\sigma(X)$ if and only if there exists a measurable $f : \mathbb{R}^n \to \mathbb{R}$ such that $Y = f(X)$.

Definition 1.12.1 Suppose $(\Omega, \mathcal{A}, P)$ is a probability space, and $\{\mathcal{G}_i\}_{i \in I}$ are sub-σ-fields of $\mathcal{A}$. These σ-fields are said to be **independent** if, for all choices of distinct $i_1, \ldots, i_k \in I$ for $k = 1, 2, \ldots$, and $A_{i_j} \in \mathcal{G}_{i_j}$,

$$P\left(\bigcap_{j=1}^k A_{i_j}\right) = \prod_{j=1}^k P(A_{i_j}).$$

Random variables $\{X_i\}$ are said to be independent if the σ-fields $\{\sigma(X_i)\}$ are independent. They are said to be **identically distributed** if each X_i has the same probability distribution. The abbreviation **iid** will be used for "independent and identically distributed". ◇

Exercise 1.12.3 Suppose $\mathcal{G}_1, \mathcal{G}_2$ and $\mathcal{G}_3$ are three independent σ-fields. Show that $\mathcal{G}_1$ and $\mathcal{G}_2 \bigvee \mathcal{G}_3$ are independent.

Exercise 1.12.4 Suppose $\{\mathcal{G}_i\}_{i \in I}$ are independent σ-fields in $(\Omega, \mathcal{G}, P)$. For every $i \in I$, let $\bar{\mathcal{G}}_i$ be the completion of $\mathcal{G}_i$. Show that $\{\bar{\mathcal{G}}_i\}_{i \in I}$ are independent.

Exercise 1.12.5 Let X, Y, Z be independent random variables. Show that, if f_1 and f_2 are measurable, then $f_1(X, Y)$ and $f_2(Z)$ are independent.

Exercise 1.12.6 Let X and Y be integrable and independent random variables. Show that XY is integrable, and $\mathrm{E}(XY) = \mathrm{E}(X)\mathrm{E}(Y)$. Now extend the result to n random variables.

Exercise 1.12.7 Let X, Y be independent random variables in L^2. Show that $X + Y \in L^2$, and

$$\mathrm{Var}(X + Y) = \mathrm{Var}(X) + \mathrm{Var}(Y).$$

Extend the result to n independent random variables.

1.13 Exercises

Exercise 1.13.1 Show by example that, a monotone class need not be a field, and a field need not be a monotone class.

Exercise 1.13.2 Let $(\Omega, \mathcal{A})$ be a measurable space. Let $\{A_i\}$ be a disjoint measurable countable partition of Ω. Show that the smallest σ-field containing $\{A_i\}$ is the collection of all sets B which can be written as the union of some sub-collection of $\{A_i\}$.

Exercise 1.13.3 Show that there does not exist any σ-field which is countably infinite.

Exercise 1.13.4 Establish the uniqueness claim in Carathéodory's extension Theorem 1.3.1.

Exercise 1.13.5 Suppose $(\Omega, \mathcal{G}, \mu)$ is a measure space where μ is a finite measure. Let $\mathcal{F}$ be a field such that $\sigma(\mathcal{F}) = \mathcal{G}$. Show that, given any $A \in \mathcal{G}$ and $\varepsilon > 0$, there is a $B \in \mathcal{F}$ such that $\mu(A - B) + \mu(B - A) < \varepsilon$.

Exercise 1.13.6 Prove the Borel–Cantelli Lemma 1.3.1.

Exercise 1.13.7 If $\{X_n\}$ converges to X in probability, show that there is a subsequence $\{n_k\}$ such that $\{X_{n_k}\}$ converges to X almost surely.

Exercise 1.13.8 Let $(\Omega, \mathcal{A}, P)$ be a probability space, and $\mathcal{G} \subset \mathcal{A}$ be a complete σ-field. Suppose $\{X_n\}$ is $\mathcal{G}$-measurable, and $X_n \xrightarrow{P} X$. Show that X is $\mathcal{G}$-measurable.

Exercise 1.13.9 Prove Part (iv) of Theorem 1.5.1, using its Part (iii) and Exercise 1.13.7.

Exercise 1.13.10 Show that d_p is a not a metric when $p < 1$.

Exercise 1.13.11 Prove the Markov, Chebyshev, and Cauchy–Schwarz inequalities.

Exercise 1.13.12 Let $\{X_i\}$ be iid random variables with finite mean μ. Let $S_n := \sum_{i=1}^{n} X_i, n \geq 1$ be the sequence of partial sums. The **Weak Law of Large Numbers** (**WLLN**) says that if $E(X_1^2) < \infty$, then $S_n/n \xrightarrow{P} \mu$. Prove this by using Markov inequality on $(S_n - n\mu)^2/n^2$.

Exercise 1.13.13 Let F be a distribution function and μ be the corresponding Lebesgue–Stieltjes measure. Show that for $a < b, a, b \in \mathbb{R}$,

$$\mu(a, b] = F(b) - F(a), \qquad \mu[a, b] = F(b) - F(a-),$$
$$\mu(a, b) = F(b-) - F(a), \qquad \mu[a, b) = F(b-) - F(a-).$$

If F is continuous at a and b, then all of the above are equal.

Exercise 1.13.14 Let F be a distribution function and let μ be the corresponding Lebesgue-Stieltjes measure. Show that

$$\mu(-\infty, b] = F(b) - F(-\infty), \qquad \mu[a, \infty) = F(\infty) - F(a-),$$
$$\mu(-\infty, b) = F(b-) - F(-\infty), \qquad \mu(a, \infty) = F(\infty) - F(a),$$
$$\mu(\mathbb{R}) = F(\infty) - F(-\infty).$$

Exercise 1.13.15 Suppose F is a distribution function and μ is the corresponding measure. Show that $\mu\{x\} = F(x) - F(x-)$, for all $x \in \mathbb{R}$. Hence for any $x \in \mathbb{R}$, $\mu\{x\} = 0$ if and only if F is continuous at x.

Exercise 1.13.16 If F is a distribution function, show that the set of its discontinuity points is countable. First assume $F(\infty) - F(-\infty) < \infty$.

Exercise 1.13.17 Suppose f is a non-negative continuous function on $\mathbb{R}$. Fix $F(0)$ arbitrarily and define

$$F(x) = \begin{cases} F(0) + \displaystyle\int_0^x f(t)dt, & \text{for } x > 0, \\ F(0) - \displaystyle\int_x^0 f(t)dt, & \text{for } x < 0. \end{cases}$$

Show that F is a distribution function. The corresponding Lebesgue–Stieltjes measure is given by

$$\mu(a, b] = \int_a^b f(t)dt, \ a, b \in \mathbb{R}, \ a < b.$$

In particular μ does not depend on the value of $F(0)$.

Exercise 1.13.18 If F is a non-decreasing, right continuous function on the interval $[a, b]$, then show that there is a unique measure μ defined on the Borel sets of $[a, b]$, such that

$$\mu\{a\} = 0 \text{ and } \mu(x, y] = F(y) - F(x) \text{ for all } a \leq x < y \leq b.$$

Exercise 1.13.19 Suppose μ is a measure. We say that it is **concentrated** on B if $\mu(\Omega \setminus B) = 0$. Suppose μ is a measure (not necessarily Lebesgue–Stieltjes) on $\mathcal{B}(\mathbb{R})$ which is concentrated on a countable set $S = \{x_1, x_2, \ldots\}$. Define

$$F(x) := \mu(-\infty, x], \ x \in \mathbb{R}.$$

Explore the properties of F.

Exercise 1.13.20 Let $S := \{x_1, x_2, \ldots\}$ be a countable subset of $\mathbb{R}$. Let μ be a measure which is concentrated on S, and $\mu\{x_i\} = a_i > 0, i \geq 1$. Then show the following:

(a) μ is a Lebesgue–Stieltjes measure if, for every finite length interval I, $\sum_{x_i \in I} a_i < \infty$.

(b) μ is a finite measure if $\sum_{i=1}^{\infty} a_i < \infty$.

(c) If μ is Lebesgue–Stieltjes, then at every $x_i \in S$, the distribution function F of μ jumps by an amount a_i. In particular, F is continuous at x if and only if $x \notin S$.

(d) If μ is a Lebesgue–Stieltjes measure and $x < y, x, y \in S$, and there is no point between x and y which is in S, then F is constant on $[x, y)$.

Exercise 1.13.21 Consider $(\mathbb{R}, \mathcal{B}(\mathbb{R}))$ and let μ be defined by

$$\mu(A) := \text{ Number of rational points in } A, \ A \in \mathcal{B}(\mathbb{R}).$$

Show that μ is a σ-finite measure but is not Lebesgue–Stieltjes, and approximation from above fails in Theorem 1.9.1.

Exercise 1.13.22 If $X_n \Rightarrow X$, then show that

$$E(|X|) \leq \liminf_{n \to \infty} E(|X_n|).$$

Exercise 1.13.23 Let $\mathbb{C}$ be the set of complex numbers. A function $f : \mathbb{R} \to \mathbb{C}$ is said to be **positive definite**, if for all $n \geq 1$, for all choices of real numbers $\{t_i\}$ and all choices of complex numbers $\{a_j\}$, we have $\sum_{i,j=1}^{n} a_i \bar{a}_j f(t_i - t_j) \geq 0$. Show that any characteristic function is positive definite. [There is a converse which says that any positive definite continuous function f such that $f(0) = 1$ is a characteristic function. See Bingham and Parthasarathy (1968) for a proof of this.]

References

R. Ash, C.A. Doléans-Dade, *Probablility & Measure Theory* (Elsevier Science, 2000)

P. Billingsley, *Probability and Measure* (John Wiley, 1995)

M.S. Bingham, K.R. Parthasarathy, A probabilistic proof of Bochner's theorem on positive definite functions. J. Lond. Math. Soc. **s1-43**(1), 626–632 (1968)

L. Breiman, *Probability* (Addison-Wesley Publishing Co., Reading, Mass.-London-Don Mills, Ont., 1968)

E. Lukacs, *Characteristic Functions* (Charles Griffin & Co., Ltd., London, 1970)

B. Ramachandran, *Advanced Theory of Characteristic Functions* (Statistical Publication Society, Calcutta, 1967)

Chapter 2
Signed Measure

We will extend the notion of a measure by dropping the condition of non-negativity. Then we will establish the first crucial technical result—the Jordan–Hahn theorem. We will then define absolute continuity of signed measures and state and prove the second crucial result—the Radon–Nikodym theorem. The concept of conditional probability and expectation, and hence of martingales, to be introduced in the next chapter, is based on this theorem. We will also define the singularity of measures and establish the Lebesgue decomposition theorem, which in particular splits any probability measure on $\mathcal{B}(\mathbb{R})$ into three unique components in terms of their relation with the Lebesgue measure.

2.1 Jordan–Hahn Decomposition

Definition 2.1.1 (*Signed measure*) Let $(\Omega, \mathcal{A})$ be a measurable space. Then $\nu : \mathcal{A} \to [-\infty, \infty]$ is said to be a **signed measure** if, $\nu(\emptyset) = 0$ and, for all countable disjoint sets $\{A_i\}$ from $\mathcal{A}$,

$$\nu\left(\bigcup_{i=1}^{\infty} A_i\right) = \sum_{i=1}^{\infty} \nu(A_i).$$

Signed measures on a field $\mathcal{F}$ are defined in the obvious way. ◇

Exercise 2.1.1 Of course every measure is a signed measure. Show that if ν is a signed measure, then it cannot take both values ∞ and $-\infty$. Show that if ν_1 and ν_2 are two measures, at least one of which is a finite measure, then $\nu = \nu_1 - \nu_2$ is a signed measure.

Definition 2.1.2 (*Indefinite integral*) Suppose $(\Omega, \mathcal{A}, \mu)$ is a measure space and f is a (real-valued) Borel measurable function on Ω such that $\int_\Omega f \, d\mu$ exists. Then

ν defined by $\nu(A) := \int_A f d\mu$, $A \in \mathcal{A}$ is called the **indefinite integral** of f with respect to μ. ◇

Exercise 2.1.2 Show that if the indefinite integral ν of f exists, then the indefinite integrals ν^+ and ν^- of f^+ and f^- are (positive) measures. Show that at least one of them is a finite measure. Moreover, $\nu = \nu^+ - \nu^-$ is a difference of two measures, and is a signed measure.

Theorem 2.1.1 Let ν be a signed measure on $(\Omega, \mathcal{A})$. Then there exists sets $C, D \in \mathcal{A}$ such that

$$\nu(C) = \sup\{\nu(A) : A \in \mathcal{A}\}, \quad \text{and} \quad \nu(D) = \inf\{\nu(A) : A \in \mathcal{A}\}.$$

♦

Example 2.1.1 (i) Suppose ν is a measure. Then $C = \Omega$, and $D = \emptyset$.

(ii) Suppose ν is the indefinite integral of f. Then it can be easily checked that we can take the sets C and D in Theorem 2.1.1 to be

$$C = \{\omega : f(\omega) \geq 0\}, \quad D = \{\omega : f(\omega) < 0\}.$$

The set $\{\omega : f(\omega) = 0\}$ could be taken out of C and included in D. Hence the pair (C, D) in Theorem 2.1.1 is not unique. ▲

For any measure, continuity from above when it is finite, and from below in general, is a standard result. We shall need the following extension to signed measures.

Exercise 2.1.3 Let ν be a countably additive set function on a σ-field $\mathcal{A}$, and $\{A_n\}_{n \geq 1} \subset \mathcal{A}$. Show that then the following hold:
(a) If $A_n \uparrow A$, then $\nu(A_n) \to \nu(A)$. The convergence may not be monotone.
(b) If $A_n \downarrow A$, and $\nu(A_i) < \infty$ for some i, then $\nu(A_n) \to \nu(A)$, need not be monotonically.
(c) The results (a) and (b) hold if ν is defined on a field $\mathcal{F}$, and we assume that $A \in \mathcal{F}$.

Proof of Theorem 2.1.1 (i) If for some $A_0 \in \mathcal{A}$, $\nu(A_0) = \infty$, then take $C = A_0$.
(ii) Suppose now that $\nu(A) < \infty$ for all $A \in \mathcal{A}$. Let

$$S := \sup\{\nu(A) : A \in \mathcal{A}\}.$$

Get $A_n \in \mathcal{A}$ such that $\nu(A_n) \to S$. Let $A_0 = \cup_{n=1}^\infty A_n$. Fix n and let $A_1^* \cap A_2^* \cap \cdots \cap A_n^*$ where each A_i^* is either A_i or $A_0 \setminus A_i$, be the 2^n disjoint sets (some sets could be empty) labeled as A_{nm}, $m = 1, 2, \ldots, 2^n$. Let

$$B_n := \cup_m \{A_{nm} : \nu(A_{nm}) \geq 0\}.$$

2.1 Jordan–Hahn Decomposition

Since each A_n is a finite disjoint union of some sets A_{nm}, and negative-valued sets have been dropped in the definition of B_n, using countable additivity of ν, we have $\nu(A_n) \leq \nu(B_n)$. There is "nesting" in the sense that if $n_1 > n_2$, then each $A_{n_1 m}$ is either a subset of $A_{n_2 m}$ or disjoint from it. This implies that for $r \geq n$,

$$\bigcup_{k=n}^{r} B_k = B_n \cup (\cup_j E_j) \text{ where, for all } j, \; E_j \cap B_n = \emptyset, \text{ and } \nu(E_j) \geq 0.$$

Hence we have

$$\nu(A_n) \leq \nu(B_n)$$
$$\leq \nu(\cup_{k=n}^{r} B_k) \text{ (additivity, and the above observation)}$$
$$\to \nu(\cup_{k=n}^{\infty} B_k) \text{ as } r \to \infty \text{ (continuity from below, Exercise 2.1.3).}$$

Let $C := \limsup B_n$. Then $\cup_{k=n}^{\infty} B_k \downarrow C$ and $0 \leq \nu(\cup_{k=n}^{\infty} B_k) < \infty$. Thus

$$S = \lim_{n \to \infty} \nu(A_n)$$
$$\leq \lim_{n \to \infty} \nu(\cup_{k=n}^{\infty} B_k)$$
$$= \nu(C) \text{ (continuity from above, Exercise 2.1.3(b))} \leq S.$$

Hence $S = \nu(C)$. The set D can be defined by considering $-\nu$. ◇

Theorem 2.1.2 (Jordan–Hahn decomposition) *Let ν be a signed measure on $(\Omega, \mathcal{A})$. Let*

$$\nu^+(A) := \sup\{\nu(B) : B \in \mathcal{A}, B \subset A\}, \; A \in \mathcal{A}$$
$$\nu^-(A) := -\inf\{\nu(B) : B \in \mathcal{A}, B \subset A\}, \; A \in \mathcal{A}.$$

Then ν^+ and ν^- are measures, and $\nu = \nu^+ - \nu^-$: ◆

Proof Since ν is a signed measure, it does not take both values $\pm\infty$. Without loss, assume that ν does not take the value $-\infty$. Let D be a set with the property described in Theorem 2.1.1. Since $\nu(\emptyset) = 0$, we have $-\infty < \nu(D) \leq 0$. Take any set $A \in \mathcal{A}$. Then

$$\nu(D) = \nu(A \cap D) + \nu(A^c \cap D).$$

Since both terms on the right side of the above equation are finite,

$$\nu(D) \leq \nu(A^c \cap D) \text{ (since } D \text{ yields the infimum)}$$
$$= \nu(D) - \nu(A \cap D).$$

This implies that
$$\nu(A \cap D) \leq 0 \text{ for any } A \in \mathcal{A}. \tag{2.1}$$

On the other hand

$$\nu(D) \leq \nu(D \cup (A \cap D^c)) \text{ (since } D \text{ yields the infimum)}$$
$$= \nu(D) + \nu(A \cap D^c) \text{ (by additivity)}.$$

This implies that
$$\nu(A \cap D^c) \geq 0 \text{ for any } A \in \mathcal{A}. \tag{2.2}$$

Take any $B \in \mathcal{A}$, $B \subset A$. Then

$$\nu(B) = \nu(B \cap D) + \nu(B \cap D^c) \text{ (by additivity)}$$
$$\leq \nu(B \cap D^c) \text{ (by (2.1))}$$
$$\leq \nu(B \cap D^c) + \nu((A \setminus B) \cap D^c) \text{ (by (2.2))}$$
$$= \nu(A \cap D^c) \text{ (by additivity)}.$$

Hence

$$\nu^+(A) \leq \nu(A \cap D^c) \text{ (taking supremum above over all } B \subseteq A)$$
$$\leq \nu^+(A) \text{ (by definition of } \nu^+).$$

This shows that
$$\nu^+(A) = \nu(A \cap D^c). \tag{2.3}$$

Similarly,

$$\nu(B) = \nu(B \cap D) + \nu(B \cap D^c) \text{ (by additivity)}$$
$$\geq \nu(B \cap D) \text{ (by (2.2))}$$
$$\geq \nu(B \cap D) + \nu((A \setminus B) \cap D) \text{ (by (2.1))}$$
$$= \nu(A \cap D) \text{ (by additivity)}.$$

Hence by taking infimum over all such B,

$$\nu^-(A) \leq -\nu(A \cap D)$$
$$\leq \nu^-(A) \text{ (by definition of } \nu^-).$$

This shows that
$$\nu^-(A) = -\nu(A \cap D). \tag{2.4}$$

Hence for all $A \in \mathcal{A}$,

$$\nu^+(A) - \nu^-(A) = \nu(A \cap D^c) + \nu(A \cap D) = \nu(A),$$

and by (2.3) and (2.4) ν^+ and ν^- are measures on $\mathcal{A}$. □

Remark 2.1.1 Let ν be a signed measure on $\mathcal{A}$ with Jordan–Hahn decomposition (ν^+, ν^-). The following statements are consequences of the arguments used in the proof of Theorem 2.1.2. Details are left as an exercise.

(a) ν is the difference of two measures, $\nu = \nu^+ - \nu^-$ and at least one of these measures is finite. Moreover, $|\nu| = \nu^+ + \nu^-$ is a measure.

(b) If $|\nu(A)| < \infty$ for all $A \in \mathcal{A}$, then $\sup_{A \in \mathcal{A}} |\nu(A)| < \infty$.

(c) There is a set $D \in \mathcal{A}$ such that for all $A \in \mathcal{A}$, $\nu(A \cap D) \leq 0$ and $\nu(A \cap D^c) \geq 0$.

(d) If $D \in \mathcal{A}$ is any set such that for all $A \in \mathcal{A}$, $\nu(A \cap D) \leq 0$ and $\nu(A \cap D^c) \geq 0$, then $\nu^+(A) = \nu(A \cap D^c)$ and $\nu^-(A) = -\nu(A \cap D)$.

(e) If E is any other set such that conditions in (d) for D are satisfied for E, then $\nu^+(D \triangle E) + \nu^-(D \triangle E) = 0$. ●

Definition 2.1.3 (*Upper, lower, and total variation*). The three measures ν^+, ν^-, and $|\nu| := \nu^+ + \nu^-$ are the **upper**, **lower**, and **total variations** of ν. ◇

Exercise 2.1.4 (*Jordan–Hahn decomposition for indefinite integral*) Let f be Borel measurable and $\int_\Omega f d\mu$ exists. Let ν by the indefinite integral of f. Show that the Jordan–Hahn decomposition of ν is given by

$$\nu^+(A) = \int_A f^+ d\mu, \quad \nu^-(A) = \int_A f^- d\mu, \quad \text{and} \quad |\nu|(A) = \int_A |f| d\mu.$$

2.2 Absolute Continuity, Radon–Nikodym Theorem

Suppose f is a non-negative measurable function on $(\Omega, \mathcal{A}, \mu)$. Recall that its indefinite integral ν is then also a measure. We may visualize the relation between μ, f, and ν as $d\nu = f d\mu$. In other words, the "derivative of ν with respect to μ equals f". We now make this and related ideas precise.

Definition 2.2.1 (*Absolute continuity*) Let μ and ν, respectively, be a measure and a signed measure on $(\Omega, \mathcal{A})$. We say that ν is **absolutely continuous** with respect to μ if, for every $A \in \mathcal{A}$, $\mu(A) = 0$ implies $\nu(A) = 0$. We write $\nu \ll \mu$. ◇

Exercise 2.2.1 Let μ and ν, respectively, be a measure and a signed measure. Show that $\nu \ll \mu$ if and only $\nu^+ \ll \mu$ and $\nu^- \ll \mu$.

Example 2.2.1 Let f be a Borel measurable function on $(\Omega, \mathcal{A}, \mu)$ such that $\int_\Omega f d\mu$ exists. If ν is the indefinite integral of f, then $\nu \ll \mu$. ▲

Exercise 2.2.2 Suppose ν is a signed measure and μ is a measure such that $\nu \ll \mu$. Show that all three measures ν^+, ν^-, and $|\nu|$ are absolutely continuous with respect to μ.

The Radon–Nikodym theorem provides a converse to the above. It is the fundamental tool to define conditional expectation, and thereby initiate the study of martingales. We shall need the following fact in its proof.

Exercise 2.2.3 Suppose $(\Omega, \mathcal{A}, \mu)$ is a measure space. Suppose f and g are Borel measurable functions. Consider the condition

$$\int_A g\,d\mu \leq \int_A f\,d\mu, \text{ for all } A \in \mathcal{A}. \tag{2.5}$$

Then show the following:

(a) If f and g are integrable and (2.5) holds, then $g \leq f$ almost surely.

(b) If integrals of f and g exists, (2.5) holds, and μ is σ-finite, then $g \leq f$ almost surely. *Hint*: Reduce to finite measure case and then use the set,

$$A_n = \{\omega : g(\omega) \geq f(\omega) + \frac{1}{n}, \ |f(\omega)| \leq n\}.$$

Theorem 2.2.1 (Radon–Nikodym theorem) *Let μ and ν, respectively, be σ-finite and signed measures on $(\Omega, \mathcal{A}, \mu)$, and $\nu \ll \mu$. Then there exists a Borel measurable function $f : \Omega \to \bar{\mathbb{R}}$ such that*

$$\nu(A) = \int_A f\,d\mu \text{ for all } A \in \mathcal{A}. \tag{2.6}$$

This f is unique: if g is any other function which satisfies (2.6), then $f = g$ almost surely μ. ◆

The function f in Theorem 2.2.1 is called the **Radon–Nikodym derivative** of ν with respect to μ, and we write (2.6) as $d\nu = f d\mu$, or $d\nu/d\mu = f$.

Proof of Theorem 2.2.1 We will prove the existence in five steps. Then the uniqueness will follow immediately from Exercise 2.2.3.

Step 1: Suppose that μ and ν are finite measures. Define

$$\mathcal{S} := \{f : f \geq 0, \text{ is } \mu \text{ integrable, and } \int_A f\,d\mu \leq \nu(A) \text{ for all } A \in \mathcal{A}\}.$$

Note that $\mathcal{S}$ is non-empty. Let

$$s := \sup\left\{\int_\Omega f\,d\mu : f \in \mathcal{S}\right\}.$$

Note that $s \leq \nu(\Omega) < \infty$. Partially order $\mathcal{S}$ by declaring that $f \geq g$ if and only if $f \geq g$ almost surely μ.

Let $f, g \in \mathcal{S}$. Then $h := \max(f, g) \in \mathcal{S}$. To see why this is true, take $B = \{\omega : f(\omega) \leq g(\omega)\}$, and observe that for any $A \in \mathcal{A}$,

2.2 Absolute Continuity, Radon–Nikodym Theorem

$$\int_A h d\mu = \int_{A \cap B} g d\mu + \int_{A \cap B^c} f d\mu$$
$$\leq \nu(A \cap B) + \nu(A \cap B^c)$$
$$= \nu(A).$$

We identify a *maximal element* of $\mathcal{S}$. Let $\{f_n\}$ be a sequence in $\mathcal{S}$ such that $\int f_n d\mu \to s$. Let $g_n := \max(f_1, \ldots, f_n)$. Then $g_n \in \mathcal{S}$ and g_n is non-decreasing. Let $g := \lim g_n$. By MCT Theorem 1.5.1(ii),

$$\int_\Omega g d\mu = \lim \int_\Omega g_n d\mu \geq \lim \int_\Omega f_n d\mu = s.$$

Let $A \in \mathcal{A}$. Then $0 \leq g_n I_A \uparrow g I_A$. Hence, by MCT,

$$\int_\Omega g I_A d\mu = \lim \int_\Omega g_n I_A d\mu.$$

But we know that $\int_\Omega g_n I_A d\mu \leq \nu(A)$ for all n. Hence $g \in \mathcal{S}$. Since $\int_\Omega g d\mu = s$, g is a maximal element of $\mathcal{S}$.

Now consider the set function

$$\nu_1(A) = \nu(A) - \int_\Omega g I_A d\mu, \quad A \in \mathcal{A}. \tag{2.7}$$

Then ν_1 is a finite measure and $\nu_1 \ll \mu$. If ν_1 is identically 0, then we are done.

Suppose if possible, ν_1 is not identically 0. Then $\nu_1(\Omega) > 0$. Hence there exists a $k > 0$, such that

$$\mu(\Omega) - k\nu_1(\Omega) < 0. \tag{2.8}$$

Using Remark 2.1.1(c), we obtain a $D \in \mathcal{A}$, such that for all $A \in \mathcal{A}$,

$$\mu(A \cap D) - k\nu_1(A \cap D) \leq 0 \text{ and } \mu(A \cap D^c) - k\nu_1(A \cap D^c) \geq 0. \tag{2.9}$$

Let if possible $\mu(D) = 0$. Then by absolute continuity, $\nu(D) = 0$. Hence using (2.7) we have $\nu_1(D) = 0$.

Using (2.9) with $A = \Omega$, we obtain

$$0 \leq \mu(D^c) - k\nu_1(D^c)$$
$$= \mu(\Omega) - k\nu_1(\Omega) \text{ (since } \mu(D) = \nu_1(D) = 0\text{)}$$
$$< 0 \text{ (by (2.8))},$$

which is a contradiction. Hence $\mu(D) > 0$. Now define

$$h(\omega) := \begin{cases} \dfrac{1}{k} & \text{if } \omega \in D, \\ 0 & \text{if } \omega \notin D. \end{cases}$$

If $A \in \mathcal{A}$, then

$$\int_A h d\mu = \frac{1}{k}\mu(A \cap D) \leq \nu_1(A \cap D) \text{ (by (2.9))}$$

$$\leq \nu_1(A) = \nu(A) - \int_A g d\mu.$$

This implies that

$$\int_A (h+g) d\mu \leq \nu(A).$$

But then $h + g > g$ on D with $\mu(D) > 0$. This contradicts the maximality of g. Thus ν_1 is identically 0, and the theorem is proved in this special case.

Step 2: Now let μ and ν be finite and σ-finite, respectively. Let $\{\Omega_n\}$ be disjoint sets in $\mathcal{A}$ such that $\cup_{n=1}^\infty \Omega_n = \Omega$, and $\nu(\Omega_n) < \infty$ for all n. Define

$$\nu_n(A) := \nu(A \cap \Omega_n), \quad A \in \mathcal{A}.$$

Then it trivially follows that for every n, ν_n is a finite measure, and $\nu_n \ll \mu$. Hence by Step 1, for every n, there exists a non-negative measurable function g_n such that ν_n is the indefinite integral of g_n with respect to μ. Take $g = \sum_{n=1}^\infty g_n$. Then ν is the indefinite integral of g with respect to μ. Note that g may be extended real valued.

Step 3: If $\nu(A) = \infty$ for all $A \in \mathcal{A}$, then take $f = \infty$. Now suppose that μ and ν are finite and arbitrary measures, respectively.

For any $C \in \mathcal{A}$, define

$$\mathcal{A}_C := \{A \cap C : A \in \mathcal{A}\}.$$

Then $\mathcal{A}_C$ is a σ-field of subsets of C.

Define the class of sets

$$\mathcal{C} := \{C \in \mathcal{A} : \nu \text{ restricted to } \mathcal{A}_C \text{ is a } \sigma\text{-finite measure}\}.$$

If for every non-empty set $A \in \mathcal{A}$, $\nu(A) = \infty$, then we can take $f \equiv \infty$. If there is a set with finite positive ν-measure, then $\mathcal{C}$ is not empty.

Let

$$s := \sup\{\mu(A) : A \in \mathcal{C}\}.$$

Pick $C_n \in \mathcal{C}$ such that $\mu(C_n) \to s$. Let $C := \cup_{n=1}^\infty C_n$. Then $C \in \mathcal{C}$. Since $s \geq \mu(C) \geq \mu(C_n) \to s$, we have $\mu(C) = s$.

Now consider the measures μ and ν restricted to $\mathcal{A}_C$ for the above choice of C. Since μ and ν are, respectively, finite and σ-finite on $\mathcal{A}_C$, by Step 2, there exists a non-negative function $f_C : C \to \bar{\mathbb{R}}$, which is measurable with respect to $\mathcal{A}_C$, and ν is the indefinite integral of μ. In other words,

2.2 Absolute Continuity, Radon–Nikodym Theorem

$$\nu(A \cap C) = \int_{A \cap C} f_C d\mu, \quad \text{for all } A \in \mathcal{A}.$$

Consider any $A \in \mathcal{A}$. Then we have either Case 1 or Case 2 given below:

Case 1. $\mu(A \cap C^c) > 0$. Then suppose if possible, $\nu(A \cap C^c) < \infty$. But this would imply that $C \cup (A \cap C^c) \in \mathcal{C}$, and

$$s \geq \mu(C \cup (A \cap C^c)) = \mu(C) + \mu(A \cap C^c) > \mu(C) = s.$$

This is a contradiction. Hence we must have $\nu(A \cap C^c) = \infty$.

Case 2. $\mu(A \cap C^c) = 0$. Then by absolute continuity, $\nu(A \cap C^c) = 0$.

Observe that in either case,

$$\nu(A \cap C^c) = \int_{A \cap C^c} \infty \, d\mu.$$

It follows that,

$$\nu(A) = \nu(A \cap C) + \nu(A \cap C^c)$$
$$= \int_A f d\mu, \quad \text{for all } A \in \mathcal{A},$$

where

$$f(\omega) = \begin{cases} f_C(\omega) & \text{if } \omega \in C, \\ \infty & \text{if } \omega \in C^c. \end{cases} \tag{2.10}$$

Clearly f is a Borel measurable function.

Step 4: Let μ and ν be σ-finite and arbitrary measures, respectively. Let Ω_n be disjoint sets in $\mathcal{A}$ such that $\Omega = \bigcup_{n=1}^{\infty} \Omega_n$ and $\mu(\Omega_n) < \infty$ for all n.

By Step 3, for every n, there exists $g_n : \Omega_n \to \bar{\mathbb{R}}$ which is Borel measurable with respect to $\mathcal{A}_{\Omega_n}$, and $\nu(A \cap \Omega_n) = \int g_n d\mu$ for every $A \in \mathcal{A}$.

Extend g_n to all of Ω by defining it to be 0 on Ω_n^c. Call this new function f_n. Note that f_n is measurable, and

$$\nu(A \cap \Omega_n) = \int_A f_n d\mu, \quad A \in \mathcal{A}.$$

Then for all $A \in \mathcal{A}$,

$$\nu(A) = \sum_{n=1}^{\infty} \nu(A \cap \Omega_n)$$

$$= \sum_{n=1}^{\infty} \int_A f_n d\mu = \int_A f d\mu \text{ where, } f := \sum_{n=1}^{\infty} f_n.$$

Step 5: Now assume that μ and ν are σ-finite and signed measures, respectively. Write $\nu = \nu^+ - \nu^-$. Without loss, assume that ν^- is a finite measure. Note that, $\nu^+ \ll \mu$ and $\nu^- \ll \mu$. By Step 4, there exist non-negative Borel measurable functions f_1 and f_2, such that ν^+ and ν^- are the indefinite integrals of f_1 and f_2 with respect to μ. Since ν^- is finite, f_2 is μ-integrable. Hence

$$\nu(A) = \nu^+(A) - \nu^-(A)$$
$$= \int_A f_1 d\mu - \int f_2 d\mu$$
$$= \int_A (f_1 - f_2) d\mu \text{ (by the additivity of integrals)}.$$

This completes the proof of the Radon–Nikodym theorem. □

Remark 2.2.1 The following facts follow from the above theorem and its proof. Details are left as exercises. Suppose $\nu \ll \mu$ where ν is a signed measure and μ is σ-finite.

(a) If ν is a finite measure then $d\nu/d\mu$ is μ-integrable, and hence is finite almost surely μ.

(b) If $|\nu|$ is σ-finite, then $d\nu/d\mu$ is finite almost surely μ.

(c) If ν is a measure, then $d\nu/d\mu \geq 0$ almost surely μ. ●

2.3 Singularity and Lebesgue Decomposition

Definition 2.3.1 (*Singularity*). Let μ_1 and μ_2 be two measures on $(\Omega, \mathcal{A})$. They are said to be **mutually singular** if there exists a set $A \in \mathcal{A}$ such that $\mu_1(A) = \mu_2(A^c) = 0$, and we write $\mu_1 \perp \mu_2$. Signed measures ν_1 and ν_2 are said to be **mutually singular** if $|\nu_1| \perp |\nu_2|$. ◇

Exercise 2.3.1 Suppose that ν is a signed measure with Jordan–Hahn decomposition $\nu = \nu^+ - \nu^-$. Then show that $\nu^+ \perp \nu^-$.

Lemma 2.3.1 *Let μ be a measure, and λ_1 and λ_2 be signed measures on $\mathcal{A}$. Then the following hold:*

(i) *If $\lambda_1 \perp \mu$ and $\lambda_2 \perp \mu$, then $\lambda_1 + \lambda_2 \perp \mu$, whenever $\lambda_1 + \lambda_2$ is well defined.*

(ii) *$\lambda_1 \ll \mu$ if and only if $|\lambda_1| \ll \mu$.*

2.3 Singularity and Lebesgue Decomposition

(iii) If $\lambda_1 \ll \mu$ and $\lambda_2 \perp \mu$, then $\lambda_1 \perp \lambda_2$.
(iv) If $\lambda_1 \ll \mu$ and $\lambda_1 \perp \mu$, then $\lambda_1 = 0$.
(v) If λ_1 is finite, then $\lambda_1 \ll \mu$ if and only if $\lim_{\mu(A)\to 0} \lambda_1(A) = 0$. ♦

Proof (i) Suppose A and B are such that

$$\mu(A) = |\lambda_1|(A^c) = 0, \text{ and } \mu(B) = |\lambda_2|(B^c) = 0.$$

Then $\mu(A \cup B) = 0$. For every $C \subset (A \cup B)^c$, $C \in \mathcal{A}$,

$$\lambda_1(C) = \lambda_2(C) = 0, \text{ and hence } |\lambda_1 + \lambda_2|((A \cup B)^c) = 0.$$

(ii) Suppose $\lambda_1 \ll \mu$ and $\mu(A) = 0$. If $\lambda_1^+(A) > 0$, then there exists a $B \subset A$ such that $\lambda_1(B) > 0$. This is a contradiction to $\lambda_1 \ll \mu$. Hence $\lambda_1^+(A) = 0$. That is $\lambda_1^+ \ll \mu$. Similarly, $\lambda_1^- \ll \mu$, and then by Part (i), $|\lambda_1| \ll \mu$. The converse is easy.

(iii) Suppose A is such that $\mu(A) = 0$ and $|\lambda_2|(A^c) = 0$. But since $\lambda_1 \ll \mu$, by Part (ii) we know that $|\lambda_1| \ll \mu$. Hence $|\lambda_1|(A) = 0$. That is $\lambda_1 \perp \lambda_2$.

(iv) By Part (iii) $\lambda_1 \perp \lambda_1$. So, pick A such that $|\lambda_1|(A) = |\lambda_1|(A^c) = 0$. Hence $|\lambda_1|(\Omega) = 0$.

(v) Suppose $\lambda_1 \ll \mu$. Suppose if possible the condition does not hold. Then there exists $\varepsilon > 0$ and sets such that

$$\mu(A_n) < 2^{-n}, \ |\lambda_1|(A_n) \geq \varepsilon.$$

Let $A := \limsup A_n$. By Lemma 1.3.1, $\mu(A) = 0$. On the other hand, $|\lambda_1|(\cup_{k=n}^\infty A_k) \geq |\lambda_1|(A_n) \geq \varepsilon$. So $|\lambda_1|(A)| = \lim_{n\to\infty} |\lambda_1|(\cup_{k=n}^\infty A_k) \geq \varepsilon$. This is a contradiction, and the proof of the lemma is complete. □

Exercise 2.3.2 Suppose f is integrable with respect to a measure μ.
(a) Show that $\lim_{n\to\infty} \int_{\{\omega: |f(\omega)| \geq n\}} f d\mu = 0$.
(b) By using (a), show that $\lim_{\mu(A)\to 0} \int_A f d\mu = 0$.

Definition 2.3.2 Let ν be a signed measure. It is called σ-finite if its total variation measure $|\nu|$ is σ-finite. ◇

Theorem 2.3.1 (Lebesgue decomposition) *Suppose μ is a measure, and ν is a σ-finite signed measure on a measure space. Then ν has a unique decomposition into $\nu = \nu_1 + \nu_2$ such that they are both signed measures, and $\nu_1 \ll \mu$, $\nu_2 \perp \mu$.* ♦

Proof (a) First suppose ν is a finite measure. Let

$$\mathcal{C} := \{A \in \mathcal{A} : \mu(A) = 0\}, \text{ and } s := \sup\{\nu(A) : A \in \mathcal{C}\} \leq \nu(\Omega) < \infty.$$

Suppose $\{A_i\}$ from $\mathcal{C}$ are such that $\nu(A_n) \to s$. Then $C = \cup_{n=1}^\infty A_n \in \mathcal{C}$ and $\nu(C) = s$. Suppose $B \in \mathcal{C}$. Note that $C \cup B \in \mathcal{C}$, and hence

$$s \geq \nu(C \cup B) = \nu(C) + \nu(B \setminus C) \geq s.$$

It follows that
$$\nu(B \setminus C) = 0.$$

Define
$$\nu_1(A) := \nu(A \setminus C), \ \nu_2(A) := \nu(A \cap C), \ A \in \mathcal{A}.$$

Then $\nu_1 \ll \mu$. To see this, suppose $\mu(B) = 0$. Then $B \in \mathcal{C}$. Hence
$$\nu_1(B) = \nu(B \setminus C) = 0.$$

Now note that $\mu(C) = 0$, and $\nu_2(C^c) = \nu(C^c \cap C) = \nu(\emptyset) = 0$. So $\nu_2 \perp \mu$. Finally, $\nu = \nu_1 + \nu_2$. Uniqueness follows by using Lemma 2.3.1(iv).

(b) Now suppose ν is a σ-finite measure. Suppose $\{A_i\}$ is a disjoint partition of Ω such that $\nu(A_n) < \infty$ for every n. Define
$$\nu_n(A) := \nu(A \cap A_n), \ A \in \mathcal{A}.$$

Then by (a), get $\{\nu_{1n}\}$, $\{\nu_{2n}\}$, such that for every n, $\nu_n = \nu_{1n} + \nu_{2n}$, and $\nu_{1n} \ll \mu$, $\nu_{2n} \perp \mu$. Adding over n (these infinite sums make sense), we get $\nu = \sum_{n=1}^{\infty} \nu_{1n} + \sum_{n=1}^{\infty} \nu_{2n} = \nu_1 + \nu_2$, where $\nu_1 \ll \mu$, and $\nu_2 \perp \mu$. Uniqueness follows by using the uniqueness proved in Part (a).

(c) Finally let ν be a σ-finite signed measure. Then use Jordan–Hahn decomposition, and apply Part (b). Uniqueness too follows from Part (b). □

2.4 Absolutely Continuous Function

Definition 2.4.1 (*Absolutely continuous function*) Let $f : [a, b] \to \mathbb{R}$. Then f is said to be **absolutely continuous** if, given any $\varepsilon > 0$, there exists a $\delta > 0$, such that for any disjoint sub-intervals (a_i, b_i) of $[a, b]$ with $\sum_{i=1}^{n}(b_i - a_i) < \delta$, we have $\sum_{i=1}^{n} |f(b_i) - f(a_i)| < \varepsilon$. ◇

Exercise 2.4.1 (a) Show that in Definition 2.4.1, it does not make any difference if we allow countable partitions instead of finite partitions.

(b) Note that if f is absolutely continuous, then it is continuous. Give examples of continuous but not absolutely continuous functions.

(c) Show that if f and g are absolutely continuous, then so is $f - g$.

Theorem 2.4.1 (*Absolutely continuous function and measure*) *Let F, G be distribution functions on $[a, b]$ with finite Lebesgue–Stieltjes measures μ_1 and μ_2. Then $\mu := \mu_1 - \mu_2 \ll \lambda$ if and only if $f := F - G$ is absolutely continuous. Here λ is the Lebesgue measure.* ◆

2.4 Absolutely Continuous Function

Proof (a) First suppose $\mu \ll \lambda$. Fix $\varepsilon > 0$. By Lemma 2.3.1(b) and (e), there exists $\delta > 0$ such that $\lambda(A) < \delta$ implies $|\mu|(A) < \varepsilon$. Let (a_i, b_i), $1 \le i \le n$ be disjoint sub-intervals of $[a, b]$, with $\sum_{i=1}^n (b_i - a_i) < \delta$. Consider the set $A := \cup_{i=1}^n (a_i, b_i]$. Then $\lambda(A) < \delta$. Hence

$$\sum_{i=1}^n |f(b_i) - f(a_i)| = \sum_{i=1}^n |\mu(a_i, b_i]| \text{ (by definition of } \mu)$$

$$\le \sum_{i=1}^n |\mu|(a_i, b_i] \text{ (since } |\mu(A)| \le |\mu|(A) \text{ for any } A)$$

$$= |\mu|(A) \text{ (since } \mu\{b_i\} = 0 \text{ for all } i)$$

$$\le \varepsilon.$$

(b) Now let f be absolutely continuous, and so continuous. For any b,

$$\mu\{b\} = \lim_{n \to \infty} \mu(b - 1/n, b] = \lim_{n \to \infty} [f(b) - f(b - 1/n)] = 0.$$

We have to show that $\lambda(A) = 0$ implies $\mu(A) = 0$. For this, fix $\varepsilon > 0$. Choose $\delta > 0$ as in the definition of absolute continuity of f.

By Theorem 1.9.1(ii),

$$\delta(A) = \inf\{\delta(V) : V \supseteq A, \ V \text{ open}\}, \ i = 1, 2, \quad (2.11)$$

where δ is any of the measures μ_1, μ_2 or λ. As finite intersection of open sets is open, using (2.11), get a single sequence $\{V_n\}$ open, $V_n \supseteq A$, such that $\lambda(V_n) \to \lambda(A) = 0$ and $\mu(V_n) \to \mu(A)$. Choose n large such that for all $k \ge n$, $\lambda(V_k) < \delta$. Write as $V_n = \cup_{i=1}^\infty (a_i, b_i)$ (disjoint union). Then

$$|\mu(V_n)| = |\sum_{i=1}^\infty \mu(a_i, b_i)|$$

$$\le \sum_{i=1}^\infty |\mu(a_i, b_i)|$$

$$= \sum_{i=1}^\infty |\mu(a_i, b_i]| \text{ (since } \mu\{b_i\} = 0)$$

$$= \sum_{i=1}^\infty |f(b_i) - f(a_i)| \le \varepsilon.$$

Since ε was arbitrary, $\lim \mu(V_n) = 0 = \mu(A)$. □

Exercise 2.4.2 The definition of absolute continuity for functions $f : \mathbb{R} \to \mathbb{R}$ is a natural extension of the definition for f defined on any interval. Suppose F and G

are **bounded** distribution functions on $\mathbb{R}$ with Lebesgue–Stieltjes measures μ_1 and μ_2. Let $f := F - G$, and $\mu := \mu_1 - \mu_2$. Show that f is absolutely continuous if and only if $\mu \ll \lambda$.

Let F be a bounded distribution function with the finite measure μ. Then the Lebesgue decomposition Theorem 2.3.1 splits μ as $\mu = \mu_1 + \mu_2$ where $\mu_1 \ll \lambda$ and $\mu_2 \perp \lambda$. The following exercise splits μ_2 further.

Exercise 2.4.3 Suppose F is a bounded distribution function on $\mathbb{R}$. Show that we can write F, uniquely up to additive constants, as $F = F_1 + F_{21} + F_{22}$ where F_1, F_{21} and F_{22} are bounded distribution functions (with measures μ_1, μ_{21} and μ_{22}), such that

(a) $\mu_1 \ll \lambda$. So F_1 is absolutely continuous.

(b) μ_{21} is discrete. So F_{21} increases only by jumps (countably many) and $\mu_{21} \perp \lambda$.

(c) F_{22} is continuous (but not absolutely continuous) and $\mu_{22} \perp \lambda$.

Hint: Define F_{21} using the jumps of $F - F_1$ and $F_{22} = F - F_1 - F_{21}$.

The measure μ_{22} or the corresponding distribution F_{22} is often called **continuous singular**. Exercise 2.5.15 provides a specific example.

2.5 Exercises

Exercise 2.5.1 Let P be a probability measures on $\mathcal{B}(\mathbb{R})$. Define a probability measure Q on $\mathcal{B}(\mathbb{R})$ by

$$Q(A) := \begin{cases} 1 & \text{if } 0 \in A, \ A \in \mathcal{B}(\mathbb{R}), \\ 0 & \text{if } 0 \notin A, \ A \in \mathcal{B}(\mathbb{R}). \end{cases} \quad (2.12)$$

Find the Jordan–Hahn decomposition of $\nu = P - Q$.

Exercise 2.5.2 (Minimality of the Jordan–Hahn decomposition) Suppose ν is a signed measure on $\mathcal{A}$. Let ν^+ and ν^- be its upper and lower variations. If $\nu = \nu_1 - \nu_2$ where ν_1 and ν_2 are measures, then show that

$$\nu_1(A) \geq \nu^+(A), \text{ and } \nu_2(A) \geq \nu^-(A), \text{ for all } A \in \mathcal{A}.$$

Exercise 2.5.3 Suppose ν is a signed measure on $\mathcal{A}$. Show that for any $A \in \mathcal{G}$, its total variation measure $|\nu|(A)$ is given by

$$|\nu|(A) = \sup\left\{\sum_{i=1}^{n} |\nu(E_i)| : \{E_i\} \text{ are disjoint measurable subsets of } A\right\}.$$

Exercise 2.5.4 Suppose ν_1 and ν_2 are signed measures. Then show that $|\nu_1 + \nu_2| \ll |\nu_1| + |\nu_2|$.

2.5 Exercises

Exercise 2.5.5 Give an example where g is finite almost surely μ, $\int g d\mu$ exists, and if ν is defined by $\nu(A) := \int_A g d\mu$, $A \in \mathcal{A}$ then none of the measures $|\nu|$ and ν are σ-finite.

Exercise 2.5.6 Give an example to show that the condition μ is σ-finite cannot be dropped from the Radon–Nikodym Theorem 2.2.1.

Exercise 2.5.7 Suppose ν is a finite signed measure, and μ is a measure. Show that $\nu \ll \mu$ if and only if given any $\varepsilon > 0$ there exists a $\delta > 0$ such that for any $A \in \mathcal{A}$, $\mu(A) < \delta$ implies $|\nu(A)| < \varepsilon$.

Exercise 2.5.8 Suppose g is a non-negative measurable function, and ν is the indefinite integral of g with respect to μ (that is for all $A \in \mathcal{A}$, $\nu(A) = \int_A g d\mu$). When is μ an indefinite integral of ν?

Exercise 2.5.9 Suppose ν_1 and ν_2 are signed measures, and $\nu_1 + \nu_2$ is well defined. Let μ be a σ-finite measure such that $\nu_i \ll \mu$, $i = 1, 2$. Show that $(\nu_1 + \nu_2) \ll \mu$, and $d(\nu_1 + \nu_2)/d\mu = (d\nu_1/d\mu) + (d\nu_2/d\mu)$.

Exercise 2.5.10 Let μ_1, μ_2, μ_3, be σ-finite measures where $\mu_1 \ll \mu_2$ and $\mu_2 \ll \mu_3$. Show that $\mu_1 \ll \mu_3$, and $d\mu_1/d\mu_3 = (d\mu_1/d\mu_2)(d\mu_2/d\mu_3)$.

Exercise 2.5.11 Let μ and ν be **mutually absolutely continuous** σ-finite measures. Show that $d\mu/d\nu = (d\nu/d\mu)^{-1}$ almost surely μ, or equivalently, almost surely ν.

Exercise 2.5.12 Suppose g is Lebesgue integrable on $\mathbb{R}$. Define

$$f(x) := \int_{(-\infty, x]} g(y)\lambda(dy), \quad x \in \mathbb{R}.$$

Show that f is absolutely continuous, and hence is continuous, on $\mathbb{R}$. *Hint*: Use DCT.

Exercise 2.5.13 Suppose g is a Lebesgue integrable function on $[a, b]$, and is continuous at $x_0 \in (a, b)$. Suppose

$$f(x) - f(a) = \int_a^x g(t)\lambda(dt), \quad a \le x \le b. \qquad (2.13)$$

Show that f is differentiable at x_0, and $f'(x_0) = g(x_0)$.

Exercise 2.5.14 (Cantor set) Start with the interval $[0, 1]$. Remove the middle $1/3$ *open* interval $E_1 = (1/3, 2/3)$. From each of the two disjoint sub-intervals of $[0, 1] \setminus E_1$, remove again their middle $1/3$ open intervals. Call their union E_2. From the four disjoint closed intervals of $[0, 1] \setminus (E_1 \cup E_2)$, remove the *four* middle $1/3$ open intervals. Continue this process of removal. Let $\{E_i\}$ be the collection of all the intervals that have been removed. Then $C := [0, 1] \setminus \bigcup_{n=1}^\infty E_n$ is called the *Cantor set*. Show that

(a) C is uncountable;

(b) $\lambda(C) = 0$;

(c) C is a closed set (in the usual topology);

(d) every point in C is a limit point;

(e) C is nowhere dense.

Exercise 2.5.15 (Cantor distribution function) In the above exercise, note that the removed set $\cup_{i=1}^{n} E_i$ consists of $2^n - 1$ disjoint intervals. Let $A_1, \ldots, A_{2^n-1}$ be their enumeration in increasing order. Define the function $F_n : [0, 1] \to [0, 1]$ by

$$F_n(x) := \begin{cases} 0 & \text{if } x = 0, \\ k/2^n & \text{if } x \in A_k, \ k = 1, 2, \ldots, 2^n - 1, \\ 1 & \text{if } x = 1. \end{cases}$$

Complete the definition by linear interpolation at other points. Show that

(a) each F_n is a non-decreasing continuous function.

(b) $F_n(x)$ converges for every x. Let $F(x) := \lim_{n \to \infty} F_n(x)$. It is called the *Cantor function*.

(c) F is continuous and non-decreasing.

(d) $F' = 0$ almost surely λ.

(e) F is not absolutely continuous.

(f) if μ is the Lebesgue–Stieltjes measure corresponding to F, then $\mu \perp \lambda$, and μ has no discrete part. That is, $\mu\{x\} = 0$ for every $x \in \mathbb{R}$.

Exercise 2.5.16 Let μ be a translation invariant measure on the Borel subsets of $\mathbb{R}^d$. Show that there exists a constant $c \in [0, \infty)$ such that $\mu(B) = c\lambda_d(B)$ for all Borel subsets B, where λ_d is the d-dimensional Lebesgue measure. For a solution see Elekes and Keleti (2006).

Reference

M. Elekes, T. Keleti, Is Lebesgue measure the only sigma-finite invariant Borel measure? Math. Anal. Appl. **321**, 445–451 (2006)

Chapter 3
Conditional Expectation

The primary notion on which the theory of martingales rests is conditional expectation. In elementary probability theory, conditional expectations are calculated given a finite number of random variables. This is based on the appropriate conditional probability distribution. More generally, we may consider calculating conditional expectation of a random variable given a σ-field. This idea is made rigorous with the help of the Radon–Nikodym theorem. We also discuss uniform integrability of random variables, which will be a crucial tool in the subsequent developments on martingales.

3.1 Conditional Expectation

Let us present two simple examples to motivate the general definition. Suppose $(\Omega, \mathcal{A}, P)$ is a probability space, and B is an event such that $P(B) > 0$. Then the conditional probability of any $A \in \mathcal{A}$ given B is defined as

$$P(A|B) := \frac{P(A \cap B)}{P(B)}.$$

Exercise 3.1.1 Show that $P(\cdot|B)$ is a valid probability measure on $(\Omega, \mathcal{A}, P)$.

Example 3.1.1 Let $\Omega := \{1, 2\} \times \{1, 2\}$ and $\mathcal{A}$ be the collection of all subsets of Ω. Let $P(1, 1) := 0.5$, $P(1, 2) := 0.1$, $P(2, 1) := 0.1$ and $P(2, 2) := 0.3$ be a probability on the above measurable space. Define the two *coordinate random variables* X and Z as $X(\omega_1, \omega_2) := \omega_1$ and $Z(\omega_1, \omega_2) := \omega_2$.

It can be easily checked that

$$P\{X = 1|Z = 1\} = \frac{P(1, 1)}{P(1, 1) + P(2, 1)} \quad (3.1)$$

$$= \frac{0.5}{0.5 + 0.1} = 5/6. \quad (3.2)$$

Likewise, $P\{X = 2|Z = 1\} = 1/6$. So, the conditional expectation of X given $\{Z = 1\}$ is

$$E(X|Z = 1) = 1 \times 5/6 + 2 \times 1/6 = 7/6.$$

Similarly

$$E(X|Z = 2) = 7/4.$$

We can combine (3.1) and (3.2) into a random variable Y:

$$Y := E(X|Z)$$
$$:= \frac{7}{6}\mathbf{1}_{\{Z=1\}} + \frac{7}{4}\mathbf{1}_{\{Z=2\}}.$$

Note that Y is a function of Z. It can be checked that $E(Y) = E(X)$. ▲

Example 3.1.2 Let X and Z be random variables on $(\Omega, \mathcal{A}, P)$, with distinct values $\{x_i\}$ and $\{z_j\}$ respectively. Further, $P(Z = z_i) > 0$ for all $i \geq 1$, and $E(|X|) < \infty$.

Then the usual conditional probability and conditional expectation of X given $Z = z_i$ are defined as

$$P\{X = x_j|Z = z_i\} := \frac{P\{X = x_j, Z = z_i\}}{P\{Z = z_i\}}, \quad j \in \{1, 2, \ldots\},$$

$$E(X|Z = z_i) := \sum_{j=1}^{\infty} x_j P\{X = x_j|Z = z_i\}, \quad i \in \{1, 2, \ldots\}.$$

We look at the above relations from the measure theoretic viewpoint. Consider the partition $A_i := Z^{-1}(z_i)$ of Ω. Then define $Y : \Omega \to \mathbb{R}$ as

$$Y(\omega) := E(X|Z = z_i), \quad \text{if } \omega \in A_i, \ i \geq 1. \quad (3.3)$$

Clearly $Y \in \sigma(Z)$, and we express (3.3) compactly as $Y = E(X|Z)$. Observe that $Y \in \sigma(Z)$. It has an additional property.

Fix $A \in \sigma(Z)$. Then $A = \cup_{i \in I} A_i$ for a countable set I. Then

$$E(Y\mathbf{1}_A) = \sum_{i \in I} E(Y\mathbf{1}_{A_i}) = \sum_{i \in I} E(X|Z = z_i)P(Z = z_i)$$

$$= \sum_{i \in I}\sum_{j=1}^{\infty} x_j P(X = x_j|Z = z_i)P(Z = z_i)$$

3.1 Conditional Expectation

$$= \sum_{j=1}^{\infty} x_j \sum_{i \in I} P(X = x_j, Z = z_i)$$

$$= \sum_{j=1}^{\infty} x_j P(\{X = x_j\} \cap A) = E(X \mathbf{1}_A).$$

So we have $E(Y\mathbf{1}_A) = E(X\mathbf{1}_A)$ for any $A \in \sigma(Z)$. ▲

In the previous example the conditioning random variable assumed only countably many values. One needs to be careful if this is not the case. The following example shows where things can go wrong.

Example 3.1.3 Let X and Y be iid Uniform $[0, 1]$ random variables. Let $Z := Y/X$. Without any loss define Z to be 0 when the divisor $X = 0$. Then $Y = 0$ if and only if $Z = 0$ almost surely. So can we claim that $P(X < x | Y = 0) = P(X < x | Z = 0)$ as we are conditioning on the same set? It is not hard to see that

$$\lim_{\varepsilon \to 0} P(X < x | Y \leq \varepsilon) = x$$

whereas,

$$\lim_{\varepsilon \to 0} P(X < x | Z \leq \varepsilon) = x^2.$$

So, formulation for $P(X < x | Y = y)$ or $E(X | Y = y)$ can be tricky. ▲

We now formalise the two properties of Y in Example 3.1.2.

Definition 3.1.1 (*Conditional expectation*) Let X be an integrable random variable on $(\Omega, \mathcal{A}, P)$. Let $\mathcal{G}$ be any sub-σ-field of $\mathcal{A}$. An integrable random variable Y is called the **conditional expectation of X given $\mathcal{G}$** if, it is measurable with respect to $\mathcal{G}$, and

$$\int_A X \, dP = \int_A Y \, dP, \text{ for all } A \in \mathcal{G}.$$

This is written as $E(X|\mathcal{G}) = Y$. ◇

It immediately follows that $E(Y) = E(X)$. The following result guarantees the existence and uniqueness of conditional expectation.

Theorem 3.1.1 (Conditional expectation: existence and uniqueness) *Suppose X is an integrable random variable on $(\Omega, \mathcal{A}, P)$ and $\mathcal{G}$ is a sub-σ-field of $\mathcal{A}$. Then there exists a $\mathcal{G}$-measurable and integrable random variable Y which satisfies*

$$\int_A X \, dP = \int_A Y \, dP, \text{ for all } A \in \mathcal{G}.$$

Further, if Y_1 and Y_2 are two choices that satisfy the above, then $Y_1 = Y_2$ almost surely. ◆

Proof Write
$$X^+ := \max\{X, 0\}, \quad X^- := \max\{-X, 0\}.$$

Then, $X^+, X^- \geq 0$, and $X = X^+ - X^-$. Clearly,

$$\int_\Omega X^+ dP \leq \int_\Omega |X| dP < \infty.$$

Define the measure ν^+ on $(\Omega, \mathcal{A})$ by

$$\nu^+(A) := \int_A X^+ dP, \ A \in \mathcal{G}.$$

Clearly, ν^+ is a finite measure and $\nu^+ \ll P$ on $(\Omega, \mathcal{G})$. By Radon–Nikodym Theorem 2.2.1, there exists a $\mathcal{G}$-measurable $Y^+ : \Omega \to [0, \infty)$ such that

$$\nu^+(A) = \int_A Y^+ dP, \ A \in \mathcal{G}.$$

We can define the variable Y^- analogously. Then it is easy to check that $Y := Y^+ - Y^-$ satisfies all the requirements. The uniqueness proof is an easy exercise. $\square$

Exercise 3.1.2 Let $(\Omega, \mathcal{A}, P)$ be a probability space and X be an integrable random variable. What are $E(X|\mathcal{G})$ for $\mathcal{G} = \mathcal{A}$ and $\mathcal{G} = \{\emptyset, \Omega\}$?

Example 3.1.4 Let X be an integrable r.v. on $(\Omega, \mathcal{A}, P)$. Let $\{A_i\}$ from $\mathcal{A}$ be a disjoint partition of Ω. Let $\mathcal{G} := \sigma(A_1, A_2, \ldots)$. Due to the structure of $\mathcal{G}$, $E(X|\mathcal{G})$, say Y, should be constant on each A_n. Moreover, it will not matter what value we assign to Y on the sets A_n for those n's for which $P(A_n) = 0$. Thus we can define $E(X|\mathcal{G})$ as

$$Y := \sum_{\{n: P(A_n) > 0\}} 1_{A_n} \frac{1}{P(A_n)} \int_{A_n} X dP.$$

To verify its correctness, first of all, by construction, Y is $\mathcal{G}$-measurable. Moreover,

$$\int_\Omega |Y| dP = \sum_{\{n: P(A_n) > 0\}} \left| \int_{A_n} X dP \right|$$
$$\leq \sum_{\{n: P(A_n) > 0\}} \int_{A_n} |X| dP$$
$$= \int_\Omega |X| dP < \infty.$$

3.1 Conditional Expectation

Fix $B \in \mathcal{G}$. Then, there exists $I \subset \mathbb{N}$ such that $B = \cup_{i \in I} A_i$. Hence,

$$\int_B Y dP = \sum_{\{i \in I : P(A_i) > 0\}} \int_{A_i} Y dP$$
$$= \sum_{\{i \in I : P(A_i) > 0\}} \int_{A_i} X dP = \int_B X dP.$$

▲

Example 3.1.5 Let X and Y be random variables on $(\Omega, \mathcal{A}, P)$ where X is integrable, and Y is discrete taking values $\{y_i\}$. Then, using Example 3.1.4, it follows that

$$E(X|Y) := E(X|\sigma(Y)) = \sum_{\{i:\, P(Y=y_i)>0\}} 1_{\{Y=y_i\}} \frac{1}{P(Y=y_i)} \int_{\{Y=y_i\}} X dP.$$

▲

Example 3.1.6 Let (X, Y) be a random vector on $(\Omega, \mathcal{A}, P)$, whose probability distribution is absolutely continuous on $\mathbb{R}^2$, with a density $f(\cdot, \cdot)$ with respect to the Lebesgue measure. Suppose that X is integrable. We wish to identify $E(X|Y)$.

By Fubini's Theorem 1.4.1,

$$f_Y(y) := \int_{-\infty}^{\infty} f(x, y)\lambda(dx), \ y \in \mathbb{R},$$

is a Borel measurable function, and is the probability density of Y.

Using the experience of Example 3.1.4, it is enough to focus only on the set

$$\{y : y \in \mathbb{R} : f_Y(y) > 0\}.$$

Since $E(|X|) = \int_{-\infty}^{\infty} \int_{-\infty}^{\infty} |x| f(x, y) dx dy < \infty$, by Fubini's Theorem 1.4.1, $\int_{-\infty}^{\infty} |x| f(x, y) dx$ is a measurable function of y, and is finite almost surely λ (Lebesgue measure).

Define

$$g(y) := \begin{cases} \dfrac{\int_{-\infty}^{\infty} x f(x,y) dx}{f_Y(y)} & \text{if } \int_{-\infty}^{\infty} |x| f(x,y) dx < \infty \text{ and } f_Y(y) > 0, \\ 0 & \text{otherwise.} \end{cases}$$

Then our claim is that $g(Y) = E(X|Y)$.

Now since $g : \mathbb{R} \to \mathbb{R}$ is a measurable function, we have $g(Y) : \Omega \to \mathbb{R}$ is also measurable. We first check that $E(|g(Y)|) < \infty$. Since

$$\lambda\left\{y \in \mathbb{R} : f_Y(y) > 0 \text{ and } \int_{-\infty}^{\infty} |x| f(x, y) dx = \infty\right\} = 0,$$

$$\begin{aligned}
E(|g(Y)|) &= \int_{-\infty}^{\infty} |g(y)| f_Y(y) \, dy \\
&= \int_{\{y: f_Y(y) > 0\}} |g(y)| f_Y(y) \, dy \\
&= \int_{\{y: f_Y(y) > 0\}} \left|\frac{1}{f_Y(y)} \int_{-\infty}^{\infty} x f(x, y) dx\right| f_Y(y) \, dy \\
&\leq \int_{\{y: f_Y(y) > 0\}} \int_{-\infty}^{\infty} |x| f(x, y) \, dx \, dy \\
&= E(|X|) < \infty.
\end{aligned}$$

Now fix $A \in \sigma(Y)$. Then, $A = Y^{-1}(B)$ for some $B \in \mathcal{B}$. Further,

$$\begin{aligned}
E(g(Y) \mathbf{1}_A) &= \int_{B \cap \{y: f_Y(y) > 0\}} g(y) f_Y(y) \, dy \\
&= \int_{B \cap \{y: f_Y(y) > 0\}} \int_{-\infty}^{\infty} x f(x, y) \, dx \, dy \\
&= E(X \mathbf{1}_{\{Y \in B\}}) = E(X \mathbf{1}_A).
\end{aligned}$$

This confirms the claim that $g(Y) = E(X|Y)$ almost surely. ▲

Theorem 3.1.2 (Tower property) *Let X be integrable, and $\mathcal{G}_1 \subset \mathcal{G}_2$ be two sub-σ-fields. Then*

$$E(X|\mathcal{G}_1) = E(E(X|\mathcal{G}_2)|\mathcal{G}_1) \quad \text{almost surely.} \tag{3.4}$$

♦

Proof Let $Y := E(X|\mathcal{G}_2)$ and $Z := E(X|\mathcal{G}_1)$.
Now, for any $A \in \mathcal{G}_1$,

$$\begin{aligned}
\int_A Z \, dP &= \int_A X \, dP \\
&= \int_A Y \, dP \quad \text{(since } A \in \mathcal{G}_2\text{)}.
\end{aligned}$$

That is, $E(Y|\mathcal{G}_2) = Z$ almost surely. This proves (3.4). □

The next theorem shows that the conditional expectation satisfies all the properties that the expectation does.

3.1 Conditional Expectation

Theorem 3.1.3 (Conditional expectation: basic properties) *Let X, Y, $\{X_n\}$ be integrable random variables on $(\Omega, \mathcal{A}, P)$, and $\mathcal{G} \subset \mathcal{A}$ be a σ-field. Suppose a and b are real numbers. Then the following hold:*

(i) If $X = a$ almost surely, then $\mathrm{E}(X|\mathcal{G}) = a$ almost surely.
(ii) $\mathrm{E}(aX + bY|\mathcal{G}) = a\mathrm{E}(X|\mathcal{G}) + b\mathrm{E}(Y|\mathcal{G})$ almost surely.
(iii) If $X \leq Y$ almost surely, then $\mathrm{E}(X|\mathcal{G}) \leq \mathrm{E}(Y|\mathcal{G})$ almost surely.
(iv) Almost surely, $|\mathrm{E}(X|\mathcal{G})| \leq \mathrm{E}(|X| \, |\mathcal{G})$.
(v) If $\lim_n X_n = X$ almost surely, $|X_n| \leq Y$, and Y is integrable, then

$$\lim_n \mathrm{E}(X_n|\mathcal{G}) = \mathrm{E}(X|\mathcal{G}) \text{ almost surely.}$$

♦

Proof Proofs of Parts (i) and (ii) are left as exercises. Part (ii) implies Part (iii) by considering $\mathrm{E}(Y - X|\mathcal{G})$. For Part (iv), using Parts (ii) and (iii), it follows that almost surely

$$\mathrm{E}(|X| \, |\mathcal{G}) \geq \mathrm{E}(X|\mathcal{G}), \text{ and}$$
$$\mathrm{E}(|X| \, |\mathcal{G}) \geq \mathrm{E}(-X|\mathcal{G}) = -\mathrm{E}(X|\mathcal{G}).$$

This completes the proof of Part (iv). To prove Part (v), define

$$Z_n := \sup_{k \geq n} |X_k - X|, \; n \geq 1.$$

Clearly $Z_n \downarrow 0$ almost surely. Parts (i)–(iii) imply that for $n \geq 1$,

$$|\mathrm{E}(X_n|\mathcal{G}) - \mathrm{E}(X|\mathcal{G})| \leq \mathrm{E}(Z_n|\mathcal{G}) \text{ almost surely.}$$

Therefore, it suffices to show that

$$\mathrm{E}(Z_n|\mathcal{G}) \to 0 \text{ almost surely.}$$

By Part (iii), $\mathrm{E}(Z_n|\mathcal{G})$ is non-increasing, with a limit, say Z. Further,

$$\int_\Omega Z \, dP \leq \int_\Omega \mathrm{E}(Z_n|\mathcal{G}) dP = \int_\Omega Z_n \, dP.$$

Note that $Z_n \leq 2Y$ and hence $\mathrm{E}(Z_n|\mathcal{G}) \leq 2Y$. Now DCT implies that

$$\lim_{n \to \infty} \int_\Omega Z_n \, dP = 0.$$

This completes the proof of Part (v). □

Example 3.1.7 Let X be an integrable random variable on $(\Omega, \mathcal{A}, P)$, and let $\mathcal{G}$ be a sub-σ-field of $\mathcal{A}$ which is independent of $\sigma(X)$. Then,

$$\mathrm{E}(X|\mathcal{G}) = \mathrm{E}(X).$$

To see this, fix any $A \in \mathcal{G}$. Then X and $\mathbf{1}_A$ are independent. The rest of the proof is left as an exercise. ▲

Definition 3.1.2 (*Conditional probability*) Let $(\Omega, \mathcal{A}, P)$ be a probability space and $\mathcal{G}$ a sub-σ-field of $\mathcal{A}$. The **conditional probability** of $A \in \mathcal{A}$ given $\mathcal{G}$ is same as $\mathrm{E}(\mathbf{1}_A|\mathcal{G})$. ◇

Exercise 3.1.3 Let $(\Omega, \mathcal{A}, P)$ be a probability space and $\mathcal{G}$ be a sub-σ-field. Let A_1 and A_2 be two sets from $\mathcal{A}$. Show that

$$P(A_1 \cup A_2|\mathcal{G}) = P(A_1|\mathcal{G}) + P(A_2|\mathcal{G}) - P(A_1 \cap A_2|\mathcal{G}) \text{ almost surely.}$$

Do you foresee any issues in taking the set function $A \to P(A|\mathcal{G})$ and treating it as a "probability measure" on $\mathcal{A}$? First consider the case where $\mathcal{A} := \sigma(\{A_i\}_{i \geq 1})$ (that is, $\mathcal{A}$ is countably generated).

Theorem 3.1.4 *Suppose that X, Y are random variables on $(\Omega, \mathcal{A}, P)$, such that X, Y and XY are integrable. Suppose that $\mathcal{G} \subset \mathcal{A}$ is a σ-field, and X is $\mathcal{G}$-measurable. Then,*

$$\mathrm{E}(XY|\mathcal{G}) = X\mathrm{E}(Y|\mathcal{G}) \text{ almost surely.}$$

◆

Proof The equality is trivial if X is a simple function. If X is integrable and $\mathcal{G}$-measurable, one can get $\mathcal{G}$-measurable simple functions s_n such that $|s_n| \leq |X|$, $s_n \to X$ almost surely. The proof now follows from the observations that $|s_n Y| \leq |XY|$, XY is integrable, and Theorem 3.1.3(v). □

Suppose X is a random variable such that $\mathrm{E}(X^2) < \infty$. It is easy to check that then $\mathrm{E}(X)$ is the minimizer of $\mathrm{E}(X - x)^2$ over all $x \in \mathbb{R}$. The conditional expectation has a similar interpretation.

Theorem 3.1.5 (Conditional expectation as a minimiser) *Let X be square integrable on $(\Omega, \mathcal{A}, P)$. Then for any sub-σ-field $\mathcal{G}$,*

$$\min_{Y \text{ is } \mathcal{G}\text{-measurable}} \mathrm{E}\big((X - Y)^2\big) = \mathrm{E}\big((X - \mathrm{E}(X|\mathcal{G}))^2\big).$$

◆

Proof Let $Z := \mathrm{E}(X|\mathcal{G})$. We first show that $\mathrm{E}(Z^2) < \infty$. Note that

$$0 \leq (X - Z)^2 = X^2 - Z^2 - 2Z(X - Z).$$

3.1 Conditional Expectation

Therefore, for all $n \geq 1$,

$$X^2 \geq X^2 \mathbf{1}_{\{|Z|\leq n\}} \geq Z^2 \mathbf{1}_{\{|Z|\leq n\}} + 2Z \mathbf{1}_{\{|Z|\leq n\}}(X - Z).$$

As a consequence,

$$\begin{aligned}
\mathrm{E}(X^2|\mathcal{G}) &\geq \mathrm{E}\big(Z^2 \mathbf{1}_{\{|Z|\leq n\}} + 2Z \mathbf{1}_{\{|Z|\leq n\}}(X - Z)\big|\mathcal{G}\big) \\
&= Z^2 \mathbf{1}_{\{|Z|\leq n\}} + 2Z \mathbf{1}_{\{|Z|\leq n\}} \mathrm{E}\left((X - Z)|\mathcal{G}\right) \text{ (by Theorem 3.1.4)} \\
&= Z^2 \mathbf{1}_{\{|Z|\leq n\}}.
\end{aligned}$$

Taking expectation on both sides, it follows that

$$\mathrm{E}\big(Z^2 \mathbf{1}_{\{|Z|\leq n\}}\big) \leq \mathrm{E}\big(\mathrm{E}(X^2|\mathcal{G})\big) = \mathrm{E}(X^2).$$

Letting $n \to \infty$, it follows that

$$\mathrm{E}(Z^2) \leq \mathrm{E}(X^2) < \infty.$$

Note that for any random variable Y with $\mathrm{E}(Y^2) = \infty$,

$$\mathrm{E}(X - Y)^2 = \infty > \mathrm{E}(X - Z)^2.$$

Therefore, it suffices to show that

$$\mathrm{E}(X - Z)^2 = \inf_{Y \text{ is } \mathcal{G}\text{-measurable: } \mathrm{E}Y^2 < \infty} \mathrm{E}\big((X - Y)^2\big).$$

For that purpose, fix a $\mathcal{G}$-measurable Y such that $\mathrm{E}(Y^2) < \infty$, and note that $(X - Z)(Z - Y)$ is integrable by the Cauchy–Schwarz inequality. Thus,

$$\begin{aligned}
\mathrm{E}(X - Y)^2 &= \mathrm{E}(X - Z)^2 + \mathrm{E}(Z - Y)^2 + 2\mathrm{E}\left((X - Z)(Z - Y)\right) \\
&\geq \mathrm{E}(X - Z)^2 + 2\mathrm{E}\left((X - Z)(Z - Y)\right) \\
&= \mathrm{E}(X - Z)^2 + 2\mathrm{E}\left((Z - Y)\mathrm{E}\left(X - Z|\mathcal{G}\right)\right) \text{ (as } Z - Y \in \mathcal{G}) \\
&= \mathrm{E}(X - Z)^2.
\end{aligned}$$

This completes the proof. □

3.2 Regular Conditional Distribution

Suppose X is a real-valued random variable defined on a probability space $(\Omega, \mathcal{A}, P)$. Let $\mathcal{G}$ be a sub-σ-field of $\mathcal{A}$. Then for every fixed $A \in \mathcal{A}$, we have defined $P(X \in A|\mathcal{G})$ as a $\mathcal{G}$-measurable function. However, this is defined only almost surely for every fixed A, and it is not clear if we can define a single version of $P(X \in A|\mathcal{G})$ that works for all $A \in \mathcal{A}$. This motivates the following definition. Recall that $\mathcal{B}(\mathbb{R})$ denotes the Borel σ-field on $\mathbb{R}$.

Definition 3.2.1 Let X be a real-valued random variable defined on $(\Omega, \mathcal{A}, P)$, and $\mathcal{G}$ be a sub-σ-field of $\mathcal{A}$. A function $\mu : \mathcal{B}(\mathbb{R}) \times \Omega \to [0, 1]$ is called a **regular conditional distribution** (RCD) of X given $\mathcal{G}$ if,

(a) for all fixed $\omega \in \Omega$, $\mu(\cdot, \omega)$ is a probability measure on $(\mathbb{R}, \mathcal{B}(\mathbb{R}))$, and

(b) for all $A \in \mathcal{B}(\mathbb{R})$, $\mu(A, \cdot)$ is a version of $P(X \in A|\mathcal{G})(\cdot)$. ◇

The following theorem guarantees the existence of an RCD.

Theorem 3.2.1 (RCD, existence) *An RCD of X given $\mathcal{G}$ exists.* ◆

The following result will be used in the proof of the theorem.

Exercise 3.2.1 Suppose that $F : \mathbb{Q} \to [0, 1]$ is a non-decreasing function such that $\lim_{n \to -\infty} F(n) = 0$ and $\lim_{n \to \infty} F(n) = 1$. Define

$$G(x) := \inf \{F(r) : r > x, r \in \mathbb{Q}\}, \quad x \in \mathbb{R}.$$

Show that G is a CDF on $\mathbb{R}$. If in addition, $\lim_{n \to \infty} F(r + n^{-1}) = F(r), r \in \mathbb{Q}$, then G is the unique CDF on $\mathbb{R}$ which agrees with F on $\mathbb{Q}$.

Proof of Theorem 3.2.1 Since $\mathbb{Q}$ is countable, choose $G : \mathbb{Q} \times \Omega \to \mathbb{R}$ such that, for all $r \in \mathbb{Q}$, $G(r, \cdot)$ is a version of $P(X \leq r|\mathcal{G})$. Then, the following properties of G follow immediately for all $r, s \in \mathbb{Q}$ almost surely.

$$G(r) \leq G(s) \text{ when } r \leq s, \tag{3.5}$$

$$0 \leq G(r) \leq 1, \tag{3.6}$$

$$\lim_{n \to \infty} G(r + n^{-1}) = G(r), \tag{3.7}$$

$$\lim_{n \to -\infty} G(n) = 0, \tag{3.8}$$

$$\lim_{n \to \infty} G(n) = 1. \tag{3.9}$$

Let $\Omega_0 \subset \Omega$ be the set on which (3.5)–(3.9) hold. Define for all $r \in \mathbb{Q}$ and $\omega \in \Omega$,

$$F(r, \omega) := \begin{cases} G(r, \omega) & \text{if } \omega \in \Omega_0, \\ \mathbf{1}_{\{r \geq 0\}} & \text{if } \omega \in \Omega_0^c. \end{cases}$$

Thus, (3.5)–(3.9) hold for every $\omega \in \Omega$, with G replaced by F.

3.2 Regular Conditional Distribution

Furthermore, $\Omega_0 \in \mathcal{G}$, and hence,

$$F(r, \omega) = P(X \le r | \mathcal{G})(\omega), \ r \in \mathbb{Q}, \ \omega \in \Omega, \quad (3.10)$$

is a valid version of the conditional expectation. Define

$$F(x, \omega) := \inf\{F(r, \omega) : r > x, r \in \mathbb{Q}\}, \ x \in \mathbb{Q}^c, \ \omega \in \Omega.$$

Then Exercise 3.2.1, along with (3.5)–(3.9) shows that for all $\omega \in \Omega$, $F(\cdot, \omega)$ is a valid CDF, and in particular is measurable.

If $x \in \mathbb{Q}^c$, and $r_n \in \mathbb{Q}$ is such that $r_n \downarrow x$, then the above shows that

$$F(r_n, \omega) \to F(x, \omega), \ \omega \in \Omega.$$

Since $F(r_n, \cdot)$ is a version of $P(X \le r_n | \mathcal{G})$, it follows that

$$F(x, \omega) = P(X \le x | \mathcal{G})(\omega), \ \omega \in \Omega. \quad (3.11)$$

For fixed $\omega \in \Omega$, let $\mu(\cdot, \omega)$ be the probability measure on $(\mathbb{R}, \mathcal{B}(\mathbb{R}))$ such that

$$\mu((a, b], \omega) = F(b, \omega) - F(a, \omega), \ -\infty < a < b < \infty.$$

It is immediate from (3.10) and (3.11) that

$$\mu((a, b], \omega) = P(a < X \le b | \mathcal{G})(\omega), \ \omega \in \Omega.$$

Now, using standard arguments as seen, for example, in the proof of Theorem 1.8.2, the above relation can be extended to claim that

$$\mu(B, \omega) = P(X \in B | \mathcal{G})(\omega), \text{ for all } B \in \mathcal{B}, \text{ and all } \omega \in \Omega.$$

This completes the proof. □

Conditional expectation of $f(X)$ can be calculated by using an RCD.

Theorem 3.2.2 *Let μ be an RCD of X given $\mathcal{G}$. For any measurable function $f : \mathbb{R} \to \mathbb{R}$ such that $f(X)$ is integrable,*

$$E(f(X)|\mathcal{G})(\omega) = \int_{-\infty}^{\infty} f(x)\mu(dx, \omega) \ \text{almost surely.} \quad (3.12)$$

◆

Proof Equation (3.12) holds if $f(X) = 1_{\{X \in B\}}$, $B \in \mathcal{B}(\mathbb{R})$. Then it holds for all simple functions, and then we can pass to limits, to claim it for all integrable functions $f(X)$. □

3.3 Jensen's Inequality

For any integrable random variable X, $E(X^2) \geq [E(X)]^2$. This is easy to show directly, or one can use the Cauchy–Schwarz inequality. Jensen's inequality extends this result and says that if ϕ is convex, and X, $\phi(X)$ are integrable, then $E\phi(X) \geq \phi(E(X))$. We state and prove a conditional version of this inequality that will be useful to us.

Theorem 3.3.1 (Jensen's inequality for conditional expectation) *Suppose ϕ is a convex function, and X, $\phi(X)$ are integrable. Then for any sub-σ-field $\mathcal{G}$, $\phi\big(E(X|\mathcal{G})\big) \leq E\big(\phi(X)|\mathcal{G}\big)$ almost surely.* ♦

Lemma 3.3.1 *Suppose $\phi : \mathbb{R} \to \mathbb{R}$ is a convex function. Then the right derivative $\phi'_+(x_0)$ exists, at all $x_0 \in \mathbb{R}$, is non-decreasing in x_0, and*

$$\phi(x_0) + (x - x_0)\phi'_+(x_0) \leq \phi(x) \text{ for all } x, x_0 \in \mathbb{R}.$$

♦

Proof Fix $x_0 \in \mathbb{R}$ and notice that if $x_0 < y < z$, then

$$\begin{aligned}\phi(y) &= \phi\left(\frac{z-y}{z-x_0}x_0 + \frac{y-x_0}{z-x_0}z\right) \\ &\leq \frac{z-y}{z-x_0}\phi(x_0) + \frac{y-x_0}{z-x_0}\phi(z) \quad (3.13) \\ &= \left[1 - \frac{y-x_0}{z-x_0}\right]\phi(x_0) + \frac{y-x_0}{z-x_0}\phi(z) \\ &= \phi(x_0) + (y-x_0)\frac{\phi(z) - \phi(x_0)}{z-x_0}.\end{aligned}$$

The above relation is same as saying that

$$x \mapsto \frac{\phi(x) - \phi(x_0)}{x - x_0}, \quad x > x_0, \quad (3.14)$$

is a non-decreasing function. A restatement of (3.13) is that

$$\frac{z-y}{z-x_0}\big[\phi(y) - \phi(x_0)\big] \leq \frac{y-x_0}{z-x_0}\big[\phi(z) - \phi(y)\big].$$

This is equivalent to

$$\frac{\phi(y) - \phi(x_0)}{y - x_0} \leq \frac{\phi(z) - \phi(y)}{z - y}, \quad x_0 < y < z.$$

For any $v < x_0 < w$, replacing x_0, y, z by v, x_0, w, it follows that

3.3 Jensen's Inequality

$$\frac{\phi(v) - \phi(x_0)}{v - x_0} \leq \frac{\phi(w) - \phi(x_0)}{w - x_0}.$$

That is, the non-decreasing function in (3.14) is bounded below.

Thus, $\phi'_+(x_0)$ exists, and in fact, for all $v < x_0$,

$$\frac{\phi(v) - \phi(x_0)}{v - x_0} \leq \phi'_+(x_0) = \inf_{x > x_0} \frac{\phi(x) - \phi(x_0)}{x - x_0}. \tag{3.15}$$

Hence, for all $x \neq x_0$, $\phi(x) - \phi(x_0) \geq (x - x_0)\phi'_+(x_0)$. This can be checked separately for the cases $x > x_0$ and $x < x_0$.

It also follows from (3.15) that, $\phi'_+(x)$ is non-decreasing. This completes the proof. □

Proof of Theorem 3.3.1 Let $Y := E(X|\mathcal{G})$. We first assume that for some finite a, b, $a \leq Y \leq b$.

Then Lemma 3.3.1 implies that

$$\phi(Y) + (X - Y)\phi'_+(Y) \leq \phi(X). \tag{3.16}$$

Since ϕ is convex, it is continuous, and hence $\phi(Y)$ takes values in the compact set $\phi([a, b])$. Moreover, since ϕ is convex, using Lemma 3.3.1, $\phi'_+(a) \leq \phi'_+(Y) \leq \phi'_+(b)$.

Thus, the left side of the inequality (3.16) is integrable. Hence,

$$E(\phi(X)|\mathcal{G}) \geq E\left[\phi(Y) + (X - Y)\phi'_+(Y)|\mathcal{G}\right]$$
$$= \phi(Y) + \phi'_+(Y)E(X - Y|\mathcal{G}) \quad \text{(since } Y \text{ is } \mathcal{G}\text{-measurable)}$$
$$= \phi(Y) \quad \text{(since } Y = E(X|\mathcal{G})),$$

which is the desired inequality (for bounded Y).

To prove the result in general, note that for $n \geq 1$, using (3.16),

$$\phi(Y)\mathbf{1}_{\{|Y| \leq n\}} + (X - Y)\phi'_+(Y)\mathbf{1}_{\{|Y| \leq n\}} \leq \phi(X)\mathbf{1}_{\{|Y| \leq n\}}.$$

Taking conditional expectation of both sides with respect to $\mathcal{G}$, and going through the same steps as earlier, it follows that

$$\phi(Y)\mathbf{1}_{\{|Y| \leq n\}} \leq \mathbf{1}_{\{|Y| \leq n\}} E(\phi(X)|\mathcal{G}).$$

Letting $n \to \infty$, the proof follows. □

The reader is invited to construct proofs for the following conditional Cauchy–Schwarz and Hölder's inequalities.

Theorem 3.3.2 (i) **Cauchy–Schwarz inequality** Let X and Y be random variables with $E(X^2 + Y^2) < \infty$. Then,

$$[E(XY|\mathcal{G})]^2 \leq E(X^2|\mathcal{G})E(Y^2|\mathcal{G}) \text{ almost surely.}$$

(ii) **Hölder's inequality** Let $X \in L^p$, $Y \in L^q$, where $p, q \in (1, \infty)$, $\frac{1}{p} + \frac{1}{q} = 1$. Then $XY \in L^1$ and

$$|E(XY|\mathcal{G})| \leq \left[E(|X|^p|\mathcal{G})\right]^{1/p} \left[E(|Y|^q|\mathcal{G})\right]^{1/q} \text{ almost surely.}$$

◆

3.4 Uniform Integrability

Suppose X is any random variable on $(\Omega, \mathcal{A}, P)$. Integrability of X is equivalent to any one of the following conditions:

(a) Given $\varepsilon > 0$, there is a $\delta > 0$ so that $P(A) < \delta$ implies $E(|X|1_A) < \varepsilon$.

(b) $\lim_{T \to \infty} \int_{\{|X|>T\}} |X| \, dP = 0$.

The following concept of integrability of an arbitrary collection of random variables will be useful.

Definition 3.4.1 (*Uniform integrability*) Let $(\Omega, \mathcal{A}, P)$ be a probability space. A family of real random variables $\{X_\alpha : \alpha \in I\}$ is said to be **uniformly integrable** (UI) if

$$\lim_{T \to \infty} \sup_{\alpha \in I} \int_{\{|X_\alpha|>T\}} |X_\alpha| \, dP = 0.$$

◇

We will not mention the index set I unless it is necessary.

Exercise 3.4.1 Show that $\{X_\alpha\}$ is UI if and only if the following two conditions hold:

(a) $\sup_{\alpha \in I} E(|X_\alpha|) < \infty$ and,

(b) given $\varepsilon > 0$, there is a $\delta > 0$ so that $P(A) < \delta \Rightarrow \sup_{\alpha \in I} E(|X_\alpha|1_A) < \varepsilon$.

Exercise 3.4.2 (a) Show that a family of finitely many integrable random variables is UI.

(b) If $\{X_\alpha\}$ and $\{Y_\alpha\}$ are UI, then show that $\{X_\alpha + Y_\alpha\}$ is UI.
(c) Show that $\sup_{\alpha \in I} E(|X_\alpha|) < \infty$, does not imply that $\{X_\alpha\}$ is UI.
(d) If Y is integrable and $|X_\alpha| \leq |Y|$ for all α, show that $\{X_\alpha\}$ is UI.

3.4 Uniform Integrability

Recall that if $\{X_n\}$ is a sequence of random variables converging to X either almost surely, in probability, or in distribution, then even if these random variables are integrable, $E(X_n) \to E(X)$ need not hold. A sufficient condition for this to hold is given by the domination condition in DCT.

We shall show below a stronger result that uses UI.

Theorem 3.4.1 *The following are equivalent.*

(i) $X_n \to X$ in L^1.

(ii) $X_n \xrightarrow{P} X$ and $\{X_n\}$ is UI. ◆

Remark 3.4.1 Theorem 3.4.1 implies that, if $X_n \xrightarrow{P} X$ and $\{X_n\}$ is UI, then $E(X_n) \to E(X)$. Exercise 3.4.2(d) shows this is a stronger statement than DCT. ●

Proof of Theorem 3.4.1 First suppose that (ii) holds. We show that then $E(|X|) < \infty$. Since $X_n \xrightarrow{P} X$, by Exercise 1.13.7, pick a sub-sequence $\{X_{n_k}\}$ such that $X_{n_k} \to X$ almost surely. Then

$$E(|X|) = E\big(\liminf_{k\to\infty} |X_{n_k}|\big)$$
$$\leq \liminf_{k\to\infty} E\big(|X_{n_k}|\big) \text{ (by Fatou's lemma)}$$
$$\leq \sup_{n\geq 1} E|X_n|$$
$$< \infty \text{ (since } \{X_n\} \text{ is UI)}.$$

Define
$$Y_n := |X_n - X|.$$

Clearly, $Y_n \xrightarrow{P} 0$ and $\{Y_n\}$ is UI. It remains to show that

$$\lim_{n\to\infty} E(Y_n) = 0.$$

Fix $\varepsilon > 0$. As $\{Y_n\}$ is UI, for some T, $\sup_n E\big(Y_n \mathbf{1}_{\{Y_n \geq T\}}\big) \leq \varepsilon$. Then,

$$\limsup_{n\to\infty} E(Y_n) \leq \limsup_{n\to\infty} E\big(Y_n \mathbf{1}_{\{Y_n < T\}}\big) + \limsup_{n\to\infty} E\big(Y_n \mathbf{1}_{\{Y_n \geq T\}}\big)$$
$$\leq \limsup_{n\to\infty} E\big(Y_n \mathbf{1}_{\{Y_n < T\}}\big) + \varepsilon = \varepsilon,$$

by Theorem 1.5.1, as $\{Y_n \mathbf{1}_{\{Y_n < T\}}\}$ is uniformly bounded and converges to 0 in probability. Since ε is arbitrary, (i) is established. Now suppose that (i) holds. Then, $X_n \xrightarrow{P} X$. Let $Y_n := |X_n - X|$. Then it suffices to show that $\{Y_n\}$ is UI. Fix $\varepsilon > 0$. By (i), for some N,

$$\sup_{n\geq N} E(Y_n) \leq \varepsilon.$$

For $i = 1, \ldots, N - 1$, there exists $T_i > 0$ with
$$E(Y_i 1_{\{Y_i > T_i\}}) \leq \varepsilon.$$

Let $T := \max\{T_1, \ldots, T_{N-1}\}$. Then it follows that
$$E(Y_n 1_{\{Y_n > T\}}) \leq \varepsilon, \ n \geq 1.$$

That is, $\{Y_n\}$ is UI. This completes the proof. $\square$

Exercise 3.4.3 If $\{X_\alpha\}$ is UI and $|Y_\alpha| \leq |X_\alpha|$ for all α, show that $\{Y_\alpha\}$ is UI.

The next result follows from Theorem 3.4.1.

Theorem 3.4.2 *If $X_n, X \in L^p$ for some $p \geq 1$, then the following are equivalent.*
(i) $X_n \to X$ in L^p.
(ii) The family $\{|X_n|^p : n \geq 1\}$ is UI and $X_n \xrightarrow{P} X$.
(iii) $X_n \xrightarrow{P} X$ and $|X_n|^p \to |X|^p$ in L^1. $\blacklozenge$

Proof *Equivalence of* (i) *and* (ii). If (i) holds, then $|X_n - X|^p \to 0$ in L^1. Then Theorem 3.4.1 implies that $X_n \xrightarrow{P} X$ and $\{|X_n - X|^p : n \geq 1\}$ is UI. Note that
$$|X_n|^p \leq 2^p \left(|X_n - X|^p + |X|^p\right).$$

Exercises 3.4.2 and 3.4.3 imply $\{|X_n - X|^p + |X|^p\}$ and $\{|X_n|^p\}$ are UI. Thus (ii) holds.

Conversely, if (ii) holds, then $|X_n - X|^p \xrightarrow{P} 0$ and
$$|X_n - X|^p \leq 2^p \left(|X_n|^p + |X|^p\right).$$

Once again, Exercises 3.4.2 and 3.4.3 imply that $\{|X_n - X|^p : n \geq 1\}$ is UI. This proves (i) after an appeal to Theorem 3.4.1. Equivalence of (ii) and (iii) is a restatement of Theorem 3.4.1. This completes the proof. $\square$

The next result is going to be useful to us later.

Theorem 3.4.3 *Let X be integrable and let $\{\mathcal{G}_\alpha\}$ be a family of sub-σ-fields. Then the collection $\{X_\alpha := E(X|\mathcal{G}_\alpha)\}$ is UI.* $\blacklozenge$

Proof For a fixed $\alpha \in I$ and $T > 0$, $\{X_\alpha > T\}$ and $\{X_\alpha < -T\}$ belong to $\mathcal{G}_\alpha$. Hence,
$$\int_{\{|X_\alpha| > T\}} |X_\alpha| \, dP = \int_{\{X_\alpha > T\}} X_\alpha \, dP - \int_{\{X_\alpha < -T\}} X_\alpha \, dP$$
$$= \int_{\{X_\alpha > T\}} X \, dP + \int_{\{X_\alpha < -T\}} (-X) \, dP$$

3.4 Uniform Integrability

$$= \int_{\{|X_\alpha|>T\}} |X|\,dP.$$

Fix $\varepsilon > 0$. By integrability of X, get $\delta > 0$ such that

$$\int_A |X|\,dP \leq \varepsilon, \text{ whenever } P(A) \leq \delta.$$

Let $T := \frac{E|X|}{\delta}$. By Markov inequality Theorem 1.6.1(i), for any $\alpha \in I$,

$$P(|X_\alpha| > T) \leq T^{-1}E|X_\alpha|$$
$$\leq T^{-1}E|X| = \delta.$$

Thus,

$$\int_{\{|X_\alpha|>T\}} |X_\alpha|\,dP \leq \int_{\{|X_\alpha|>T\}} |X|\,dP \leq \varepsilon.$$

This completes the proof. □

We end the section with a characterization of UI random variables.

Definition 3.4.2 (*Test function of UI*) Let $G : [0, \infty) \to [0, \infty)$ be measurable. It is called a **test function of uniform integrability** if

$$\lim_{x\to\infty} \frac{G(x)}{x} = \infty.$$

◇

Theorem 3.4.4 (Characterization of UI via test functions) *A nonempty family $\mathcal{X}$ of integrable functions is uniformly integrable if and only if there exists a test function of uniform integrability G such that*

$$\sup_{X \in \mathcal{X}} E(G(|X|)) < \infty. \tag{3.17}$$

Moreover, if G exists, it can be chosen to be non-decreasing convex. ♦

To prove the result, we need the following lemma.

Lemma 3.4.1 *Let $f : [0, \infty) \to [0, \infty)$ be a non-increasing measurable function with $f(x) \to 0$, as $x \to \infty$. Then, there exists a continuous function $g : [0, \infty) \to (0, \infty)$ such that*

$$\int_0^\infty g(x)\lambda(dx) = +\infty \text{ but } \int_0^\infty f(x)g(x)\lambda(dx) < \infty. \tag{3.18}$$

Further, g can be chosen so that $x \mapsto x \int_0^x g(\xi)\lambda(d\xi)$ is convex. ♦

Proof Suppose $\tilde{f}$ is a strictly positive continuously differentiable with $\tilde{f}(x) \geq f(x)$, for all $x \geq 0$, with $\tilde{f}(x) \to 0$, as $x \to \infty$. With such an $\tilde{f}$ and $g = -\tilde{f}'/\tilde{f}$, we have

$$\int_0^\infty g(x)\lambda(dx) = \lim_{x \to \infty} (\ln(\tilde{f}(0)) - \ln(\tilde{f}(x))) = \infty.$$

On the other hand

$$\int_0^\infty f(x)g(x)\lambda(dx) \leq \int_0^\infty \tilde{f}(x)g(x)\lambda(dx)$$
$$= \lim_{x \to \infty} (\tilde{f}(0) - \tilde{f}(x)) = \tilde{f}(0) < \infty.$$

□

Exercise 3.4.4 Show that in the above proof, $\tilde{f}$ can be constructed such that $x \mapsto -x \ln \tilde{f}(x)$ is convex.

Proof of Theorem 3.4.4 First suppose (3.17) holds for some test function of uniform integrability and the supremum equals M ($0 \leq M < \infty$). For $n > 0$, let $C_n \in \mathbb{R}$ be such that $G(x) \geq nMx$, for $x \geq C_n$. Therefore,

$$M \geq \mathrm{E}(G(|X|)) \geq \mathrm{E}\big(G(|X|)\mathbf{1}_{\{|X| \geq C_n\}}\big) \geq nM\mathrm{E}\big(|X|\mathbf{1}_{\{|X| \geq C_n\}}\big),$$

for all $X \in \mathcal{X}$. Hence, $\sup_{X \in \mathcal{X}} \mathrm{E}(|X|\mathbf{1}_{\{|X| \geq C_n\}}) \leq \frac{1}{n}$, and so $\mathcal{X}$ is UI. Conversely, suppose $\mathcal{X}$ is uniformly integrable. Let

$$f(x) = \sup_{X \in \mathcal{X}} \mathrm{E}\big(|X|\mathbf{1}_{\{|X| \geq x\}}\big), \quad x \geq 0.$$

Then f satisfies the conditions of Lemma 3.4.1, and a function g for which (3.18) holds can be constructed. Consequently, the function $G(x) = x \int_0^x g(\xi)\lambda(d\xi)$ is a test function of uniform integrability. On the other hand, for $X \in \mathcal{X}$, we have

$$\mathrm{E}(G(|X|)) = \mathrm{E}\bigg(|X| \int_0^\infty \mathbf{1}_{\{x \leq |X|\}} g(x)\lambda(dx)\bigg)$$
$$= \int_0^\infty g(x)\mathrm{E}\big(|X|\mathbf{1}_{\{|X| \leq x\}}\big)\lambda(dx)$$
$$\leq \int_0^\infty g(x)f(x)\lambda(dx) < \infty.$$

□

3.5 Exercises

Exercise 3.5.1 Suppose $\mathcal{G} = \sigma(A_1, A_2, \ldots, A_n)$, where $\{A_i\} \subset \mathcal{A}$ but not necessarily disjoint and let X be an integrable random variable. Describe $\mathrm{E}(X|\mathcal{G})$.

Exercise 3.5.2 Show that if ϕ is a convex function on any open interval in $\mathbb{R}$, then it is continuous on that interval.

Exercise 3.5.3 Let $(\Omega, \mathcal{A}, P)$ be a probability space, and $X : \Omega \to [0, \infty)$ be an $\mathcal{A}$-measurable function with $\int X\, dP = \infty$. Suppose $\mathcal{G}$ is a sub-σ-field of $\mathcal{A}$. Prove or give counter-example to the following statements.

(a) There exists a $\mathcal{G}$-measurable function $Y : \Omega \to [0, \infty)$ such that

$$\int_A X\, dP = \int_A Y\, dP, \quad \text{for all } A \in \mathcal{G}.$$

(b) There exists a $\mathcal{G}$-measurable function $Y : \Omega \to [0, \infty]$ such that

$$\int_A X\, dP = \int_A Y\, dP, \quad \text{for all } A \in \mathcal{G}.$$

Exercise 3.5.4 Let $X \in L^2(\Omega, \mathcal{A}, P)$ and $\mathcal{H} \subset \mathcal{G} \subset \mathcal{A}$ be σ-fields. Show that the covariance between $\mathrm{E}(X|\mathcal{H})$ and $\mathrm{E}(X|\mathcal{G}) - \mathrm{E}(X|\mathcal{H})$ is 0.

Exercise 3.5.5 Suppose $X \in L^2(\Omega, \mathcal{A}, P)$ and $\mathcal{G} \subset \mathcal{A}$ is a σ-field.
(a) Show that
$$\mathrm{Cov}\Big(X, \mathrm{E}(X|\mathcal{G})\Big) \geq 0.$$

(b) Show that equality holds above if and only if $\mathrm{E}(X|\mathcal{G}) = \mathrm{E}(X)$ almost surely.

Exercise 3.5.6 Suppose $\mathcal{G}_1 \subset \mathcal{G}_2$ are sub-σ-fields and $\mathrm{E}(X^2) < \infty$. Then show that

$$\mathrm{E}\Big((X - \mathrm{E}(X|\mathcal{G}_2))^2\Big) \leq \mathrm{E}\Big((X - \mathrm{E}(X|\mathcal{G}_1))^2\Big).$$

Exercise 3.5.7 Let $\{X_n\}, X$ be integrable random variables, $X_n \downarrow X$ almost surely. Show that on $\cup_{n=1}^\infty \{\mathrm{E}(X_n|\mathcal{G}) < \infty\}$, $\mathrm{E}(X_n|\mathcal{G}) \downarrow \mathrm{E}(X|\mathcal{G})$ almost surely.

Exercise 3.5.8 Suppose X and Y are random variables with finite second moments, and $\mathrm{E}(X|Y) = f(Y)$ for some decreasing function f. Show that $\mathrm{Cov}(X, Y) \leq 0$.

Exercise 3.5.9 Show that for any $a > 0$, $P(|X| \geq a|\mathcal{G}) \leq a^{-2}\mathrm{E}(X^2|\mathcal{G})$, almost surely.

Exercise 3.5.10 Let $X \in L^2(\Omega, \mathcal{A}, P)$ and $\mathcal{G}$ be a sub-σ-field of $\mathcal{A}$. Show that

$$\text{Var}\,(\text{E}(X|\mathcal{G})) \leq \text{Var}\,(X) \,. \tag{3.19}$$

Using (3.19), deduce the Rao–Blackwell theorem: *Suppose $\{P_\theta\}$ is a family of probability distributions such that $P_\theta \ll \mu$ for all θ, for some fixed measure μ and for which T is a sufficient statistic. Let X be any unbiased estimator of $f(\theta)$, so that $\text{E}_\theta(X) = f(\theta))$, with finite variance. Then,*

$$\text{E}\big[\,(\text{E}(X|\sigma(T)) - f(\theta))^2\,\big] \leq \text{E}\big[(X - f(\theta))^2\big].$$

Exercise 3.5.11 Let $X \in L^2(\Omega, \mathcal{A}, P)$ and $\mathcal{G} \subset \mathcal{A}$ be a σ-field. Let $Z := \text{E}(X|\mathcal{G})$. Suppose $Y \in L^2(\Omega, \mathcal{G}, P)$ is such that

$$\text{E}\big[(X - Y)^2\big] = \text{E}\big[(X - Z)^2\big].$$

Show that $Y = Z$ almost surely.

Exercise 3.5.12 Let $X \sim U(0, 1)$. Fix a positive integer k. Compute $\text{E}(X|Y)$ where $Y = kX - [kX]$ and $[x]$ denotes the integer part of x.

Exercise 3.5.13 Let (X, Y) be a bivariate Gaussian random variable where X and Y have 0 means, and $\text{E}(Y^2) > 0$. Show that $\text{E}(X|Y) = \rho Y$ where $\rho = \text{E}(XY)/\text{E}(Y^2)$.

Exercise 3.5.14 Let X and Y be independent random variables, and $f : \mathbb{R}^2 \to \mathbb{R}$ be a Borel measurable bounded function. Define $g : \mathbb{R} \to \mathbb{R}$ by

$$g(y) := \int_{-\infty}^{\infty} f(x, y)\, P_X(dx), \ y \in \mathbb{R}.$$

Show that $\text{E}\,(f(X, Y)|Y) = g(Y)$ almost surely.

Exercise 3.5.15 Prove or give counter-examples to the following.
(a) If X is integrable and is independent of the σ-field $\mathcal{G}$, then for any σ-field $\mathcal{H}$,

$$\text{E}\,(X|\mathcal{G} \vee \mathcal{H}) = \text{E}\,(X|\mathcal{H}) \ \text{ almost surely}\,.$$

(b) Let X be an integrable random variable and $\mathcal{G}, \mathcal{H}$ be σ-fields such that $\mathcal{H} \vee \sigma(X)$ is independent of $\mathcal{G}$. Then

$$\text{E}\,(X|\mathcal{G} \vee \mathcal{H}) = \text{E}\,(X|\mathcal{H}) \ \text{ almost surely}.$$

Exercise 3.5.16 Let X be an integrable random variable defined on $(\Omega, \mathcal{A}, P)$. Suppose that for $n \geq 1$, $\{\mathcal{G}_n\}$ is a sequence of σ-fields such that $\mathcal{G}_1 \subset \mathcal{G}_2 \subset \cdots \subset \mathcal{A}$. Suppose Y is such that

$$\text{E}(X|\mathcal{G}_n) = Y \ \text{ almost surely, for all } \ n \geq 1\,.$$

3.5 Exercises

Show that

$$E\left(X \mid \bigvee_{n=1}^{\infty} \mathcal{G}_n\right) = Y \text{ almost surely.}$$

Exercise 3.5.17 Suppose that X and Y are integrable random variables defined on a probability space $(\Omega, \mathcal{A}, P)$ such that, almost surely

$$E(X|\sigma(Y)) = Y \text{ and } E(Y|\sigma(X)) = X.$$

Show that $X = Y$ almost surely. *Hint*: First show $\int_{[Y \leq c]} (X - Y) dP = 0$ for $c \in \mathbb{R}$. Then show that $\int_{[X \leq c, Y \leq c]} (X - Y) dP \leq 0$.

Exercise 3.5.18 For any square integrable X, and any sub-σ-field $\mathcal{G}$, show that

$$\text{Var}(X) = \text{Var}(E(X|\mathcal{G})) + E((X - E(X|\mathcal{G}))^2).$$

Exercise 3.5.19 Suppose X is an integrable random variable. Show that $E(X|\mathcal{G}) = E(X)$ whenever $\mathcal{G}$ is a trivial σ-field.

Exercise 3.5.20 Suppose that X, Y are integrable, Y is measurable with respect to a sub-σ-field $\mathcal{G}$, $X - Y$ is independent of $\mathcal{G}$, and $E(X) = E(Y)$. Then show that $E(X|\mathcal{G}) = Y$.

Exercise 3.5.21 Let Y be a bounded real-valued r.v., and let $\mathcal{F} \subseteq \mathcal{G}$ be σ-fields. If $E(Y|\mathcal{F}) \stackrel{d}{=} E(Y|\mathcal{G})$, show that $E(Y|\mathcal{F}) = E(Y|\mathcal{G})$ a.s.

Exercise 3.5.22 Let X be a random variable on $(\Omega, \mathcal{A}, P)$ and $\mathcal{G}$ be a sub-σ-field of $\mathcal{A}$. If μ is the RCD of X given $\mathcal{G}$, Y is a $\mathcal{G}$-measurable random variable, and $\phi : \mathbb{R}^2 \to \mathbb{R}$ is a bounded measurable function.
Show that

$$(E(\phi(X, Y)|\mathcal{G}))(\omega) = \int_{\mathbb{R}} \phi(x, Y(\omega)) \mu(dx, \omega) \text{ for almost all } \omega.$$

Exercise 3.5.23 Let $\{X_\alpha\}$ be a collection of random variables. Show that for $p > 0$, $\{|X_\alpha|^p\}$ is UI, if for some $\varepsilon > 0$, $\sup_{\alpha \in I} E(|X_\alpha|^{p+\varepsilon}) < \infty$.

Exercise 3.5.24 Let $\{X_\alpha\}$ be a collection of random variables. Then show the following:
(a) If $|X_\alpha| \leq Y, \alpha \in I$, where Y is an integrable random variable, then $\{X_\alpha\}$ is UI.
(b) Suppose for a finite constant c, and an integrable Y,

$$\sup_{\alpha \in I} P(|X_\alpha| > x) \leq cP(|Y| > x), \text{ for all } x.$$

Then $\{X_\alpha\}$ is UI.

Exercise 3.5.25 Let $\{X_i\}$ be iid random variables with finite mean μ. Let $S_n := \sum_{i=1}^{n} X_i, n \geq 1$ be the sequence of partial sums.
(a) Show that $\{S_n/n\}$ is UI.
(b) The **Strong Law of Large Numbers (SLLN)** says that $S_n/n \to \mu$ almost surely. See Billingsley (1995) for a proof. We shall give a proof later as an application of some general result for martingales. Using (a) and SLLN, show that $n^{-1}S_n \to \mu$ in L^1.

Exercise 3.5.26 If X_n, X are in L^p for $p \geq 1$, and $X_n \xrightarrow{P} X$ as $n \to \infty$, then show that the following are equivalent.
(a) $X_n \to X$ in L^p.
(b) $|X_n| \to |X|$ in L^p.
(c) $\limsup_{n \to \infty} \mathrm{E}(|X_n|^p) \leq \mathrm{E}(|X|^p)$.
(d) $|X_n|^p \to |X|^p$ in L^1.
(e) The family $\{|X_n|^p\}$ is UI.
Hint: Show (a)$\Rightarrow$(b)$\Rightarrow$(c)$\Rightarrow$(d)$\Rightarrow$(e)$\Rightarrow$(a). For showing (c)$\Rightarrow$(d), write

$$\big| |X_n|^p - |X|^p \big| = |X_n|^p + |X|^p - 2\left(|X_n|^p \wedge |X|^p\right).$$

Reference

P. Billingsley, *Probability and Measure*, 3rd edn (John Wiley, 1995)

Chapter 4
Martingales

Consider a player who is playing a game that progresses sequentially over time. At any time instant, the player's strategy depends on the available accumulated information till that time. We could call this game fair if, the player's accumulated gain at any time, on the average, is neither more nor less than what it was before that instant. This idea of a fair game will be made precise and lead us to the definition of a martingale.

Lévy (1937) introduced the concept of martingale in probability theory, even though the term "martingale" was introduced by Ville (1939). The extensive works of Doob (1984) developed the theory of martingales, and many of his results will be discussed in this book.

4.1 Definition and Examples

Definition 4.1.1 (*Filtration, adapted sequence*) Let $(\Omega, \mathcal{A}, P)$ be a probability space.

(a) A sequence of sub-σ-fields $\{\mathcal{G}_n\}_{n \geq 0}$ of $\mathcal{A}$ is said to be a **filtration** if $\mathcal{G}_0 \subset \mathcal{G}_1 \subset \cdots \subset \mathcal{A}$. Then $(\Omega, \mathcal{A}, \{\mathcal{G}_n\}_{n \geq 0}, P)$ is called a **filtered probability space**.

(b) Let $\{X_n\}_{n \geq 0}$ be random variables. We say that $(X_n, \mathcal{G}_n)$ is an **adapted** sequence or $\{X_n\}$ is adapted to $\{\mathcal{G}_n\}$ if, $\{\mathcal{G}_n\}$ is a filtration, and X_n is $\mathcal{G}_n$-measurable for $n \geq 0$. ◇

For any random variables $\{X_n\}$, $(X_n, \sigma(X_0, \ldots X_n))$ is always an adapted sequence. In general, $\mathcal{G}_n$ is the accumulated information for the player till time n. The player's accumulated gain X_n at time n is then $\mathcal{G}_n$-measurable. We then have the following definition.

Definition 4.1.2 (*Martingale*) Suppose $(X_n, \mathcal{G}_n)_{n \in \mathbb{N}}$ is adapted. Then $\{X_n\}$ is said to be a **martingale** with respect to the filtration $\{\mathcal{G}_n\}$ if for all $n \in \mathbb{N}$,

(a) $E(|X_n|) < \infty$, and

(b) $E(X_{n+1}|\mathcal{G}_n) = X_n$ almost surely.

We also simply say that $(X_n, \mathcal{G}_n)$ is a martingale. It is called a **sub-martingale** or a **super-martingale** according as (b) is replaced respectively by $E(X_{n+1}|\mathcal{G}_n) \geq X_n$ or $E(X_{n+1}|\mathcal{G}_n) \leq X_n$ for all n. ◇

Note that an integrable sequence $\{X_n\}$ can be a martingale with respect to more than one filtration. In any case, if $\{X_n\}$ is a martingale then $E(X_n)$ is constant over n.

Exercise 4.1.1 Suppose $(X_n, \mathcal{G}_n)_{n \geq 0}$ is a sub-martingale. Show that it is a martingale if and only if $E(X_n) = E(X_0)$ for all n.

Exercise 4.1.2 Suppose $(X_n, \mathcal{G}_n)$ is a martingale. Show that $(X_n, \bar{\mathcal{G}}_n)$ is also a martingale where $\bar{\mathcal{G}}_n$ is the completion of $\mathcal{G}_n$. Likewise for sub- and super-martingales.

Exercise 4.1.3 (a) Suppose $(X_n, \mathcal{G}_n)$ is a martingale. Then show that $(X_n, \sigma(X_1, \ldots X_n))$ is also a martingale. Note that $\{\sigma(X_0, \ldots X_n)\}$ is the smallest filtration for which $\{X_n\}$ is a martingale. Likewise for sub- and super-martingales.

(b) If $(X_n, \mathcal{G}_n)$ is a martingale and $\phi : \mathbb{R} \to \mathbb{R}$ is convex such that $\{\phi(X_n)\}$ is integrable, then show that $(\phi(X_n), \mathcal{G}_n)$ is a sub-martingale.

(c) If $(X_n, \mathcal{G}_n)$ is a sub-martingale, $\phi : \mathbb{R} \to \mathbb{R}$ is non-decreasing and convex such that $\{\phi(X_n)\}$ is integrable, then show that $(\phi(X_n), \mathcal{G}_n)$ is a sub-martingale.

The following relation between martingales and sub-martingales (or super-martingales) is easy in our set up but the analogous result for continuous time is non-trivial.

Theorem 4.1.1 (Doob's decomposition theorem) *Let $(X_n, \mathcal{G}_n)$, $n \geq 1$ be a sub-martingale. Then, X_n can be decomposed as*

$$X_n = Y_n + A_n,$$

where

(i) $(Y_n, \mathcal{G}_n)$ is a martingale,

(ii) $A_{n+1} \geq A_n$ a.s. for all $n \geq 1$,

(iii) $A_1 = 0$ a.s.

(iv) A_n is $\mathcal{G}_{n-1}$-measurable for all $n \geq 2$.

Further, Y_n and A_n are uniquely determined.

A similar decomposition holds for super-martingales. ◆

For a sub-martingale $(X_n, \mathcal{G}_n)$, the pair $\{(Y_n, A_n)\}$ given in Theorem 4.1.1 is called its **Doob decomposition**, and the sequence $\{A_n\}$ is called the **compensator** of $\{X_n\}$.

Proof Let $A_1 = 0$, and for $n \geq 2$, inductively define

$$A_n = A_{n-1} + E(X_n - X_{n-1}|\mathcal{G}_{n-1}),$$

4.1 Definition and Examples

$$Y_n = X_n - A_n.$$

Clearly, (iii) and (iv) hold, and hence, for $n \geq 2$,

$$\begin{aligned}
E(Y_n|\mathcal{G}_{n-1}) &= E(X_n|\mathcal{G}_{n-1}) - A_n \\
&= E(X_n|\mathcal{G}_{n-1}) - A_{n-1} - E(X_n - X_{n-1}|\mathcal{G}_{n-1}) \\
&= X_{n-1} - A_{n-1} \\
&= Y_{n-1},
\end{aligned}$$

showing that (i) holds.

Next observe that

$$A_n - A_{n-1} = E(X_n|\mathcal{G}_{n-1}) - X_{n-1} \geq 0 \text{ a.s.},$$

showing that (ii) holds.

Now suppose that $X_n = Z_n + B_n$ is a decomposition such that (i)–(iv) hold with A_n and Y_n replaced by B_n and Z_n respectively.

Then,

$$Y_n - Z_n = B_n - A_n, \; n \geq 1.$$

From (iv) it follows that for $n \geq 2$,

$$\begin{aligned}
B_n - A_n &= E(B_n - A_n|\mathcal{G}_{n-1}) \\
&= E(Y_n - Z_n|\mathcal{G}_{n-1}) \\
&= Y_{n-1} - Z_{n-1} \text{ (by (i))} \\
&= B_{n-1} - A_{n-1}.
\end{aligned}$$

Using (iii), it follows inductively that $Y_n = Z_n$ a.s. for all n, and the uniqueness follows. □

Example 4.1.1 (*Random walk*) Let $\{\xi_n\}$ be independent random variables with finite means $\{\mu_n\}$. Let $\mathcal{G}_n := \sigma(\xi_1, \ldots, \xi_n), n \geq 1$. Let $\mathcal{G}_0 := \{\emptyset, \Omega\}$ and $S_0 := 0$. For every $n \geq 1$, let $S_n := \xi_1 + \cdots + \xi_n$. Then $(S_n, \mathcal{G}_n)$ is adapted, and each S_n has finite mean. Using the facts that S_n is measurable with respect to $\mathcal{G}_n$, and ξ_{n+1} is independent of $\mathcal{G}_n$,

$$\begin{aligned}
E(S_{n+1}|\mathcal{G}_n) &= E(S_n + \xi_{n+1}|\mathcal{G}_n) \\
&= S_n + \mu_{n+1}.
\end{aligned}$$

If μ_n is zero for all n, then $(S_n, \mathcal{G}_n)$ is a martingale.

If $\mu_n \geq 0$ for all n, then $(S_n, \mathcal{G}_n)$ is a sub-martingale. If $\mu_n \leq 0$ for all n, then $(S_n, \mathcal{G}_n)$ is a super-martingale. In any case the sequence $(S_n - (\mu_1 + \cdots + \mu_n), \mathcal{G}_n)$ is a martingale.

If $\{\xi_n\}$ are iid then $\{S_n\}$ is called a **random walk**. The walk is called **simple symmetric** if $P(\xi_n = 1) = P(\xi_n = -1) = 1/2$, ▲

Exercise 4.1.4 Let $\{X_i\}$ be iid with $E(X_i) = 0$ and $E(X_i^2) = 1$. Let $S_n := X_1 + \cdots + X_n$ and $\mathcal{G}_n := \sigma(X_1, \ldots X_n)$. Show that $(S_n^2, \mathcal{G}_n)$ is a sub-martingale, and $(S_n^2 - n, \mathcal{G}_n)$ is a zero mean martingale.

Exercise 4.1.5 Suppose $\{X_i\}$ are independent and $1 \leq E(e^{X_i}) < \infty$ for all i. Let

$$m_n = \prod_{i=1}^n E(e^{X_i}), \quad S_n = X_1 + \cdots + X_n, \quad \mathcal{G}_n := \sigma(X_1, \ldots, X_n), \quad n \geq 1.$$

Show that $(e^{S_n}, \mathcal{G}_n)$ is a sub-martingale and $(m_n^{-1} e^{S_n}, \mathcal{G}_n)$ is a martingale.

Example 4.1.2 (*Doubling game*) Suppose a gambler starts with capital X_0. At each time $n+1$, $n \geq 0$, the gambler stakes the entire asset X_n, and then either doubles her asset or loses everything, each with probability $1/2$.

Thus, given $(X_1, \ldots, X_n)$, X_{n+1} is given by

$$X_{n+1} = \begin{cases} 2X_n & \text{with probability } 1/2, \\ 0 & \text{with probability } 1/2. \end{cases}$$

Then, $(X_n, \sigma(X_0, \ldots, X_n))$ is a martingale. ▲

Exercise 4.1.6 (*Product martingale*) Suppose $\{\xi_n\}$ are independent positive random variables with finite means $\{m_n\}$. Let $X_0 = 1$ and

$$X_n := \xi_1 \xi_2 \cdots \xi_n \text{ and } \mathcal{G}_n := \sigma(\xi_1, \ldots, \xi_n), \text{ for } n \geq 1.$$

Then show that $(M_n, \mathcal{G}_n)$ is a martingale, where

$$M_n := \frac{X_n}{m_1 \cdots m_n}.$$

Example 4.1.3 $(N_t)_{t \geq 0}$ is said to be a **Poisson process** with mean λ, if for all $k \geq 1$ and all $t_0 < t_1 < \cdots < t_k$, $N_{t_j} - N_{t_{j-1}}$, $j \leq k$ are independent, Poisson variables with parameters $\lambda(t_i - t_{j-1})$ respectively. Let $X_n := N_n - n\lambda$, $\mathcal{G}_n := \sigma(N_t : t \leq n)$, $n \geq 1$. Then

$$\begin{aligned} E(X_{n+1}|\mathcal{G}_n) &= E(N_{n+1} - (n+1)\lambda|\mathcal{G}_n) \\ &= E(N_{n+1} - N_n|\mathcal{G}_n) + N_n - (n+1)\lambda \\ &= \underbrace{E(N_{n+1} - N_n)}_{=\lambda} + N_n - (n+1)\lambda \\ &= N_n - n\lambda. \end{aligned}$$

Hence $(X_n, \mathcal{G}_n)$ is a martingale. ▲

4.1 Definition and Examples

Example 4.1.4 (*Doob's martingale*) Let Z be an integrable r.v. on $(\Omega, \mathcal{A}, P)$. Then $(X_n := \mathrm{E}(Z|\mathcal{G}_n), \mathcal{G}_n)$ is a martingale for any filtration $\{\mathcal{G}_n\}$. To check this, first observe that for each n, X_n is $\mathcal{G}_n$-measurable and integrable. Now recall the tower property Theorem 3.1.2.

Using that, for every $n \geq 0$,

$$\mathrm{E}(X_{n+1}|\mathcal{G}_n) = \mathrm{E}\big(\mathrm{E}(Z|\mathcal{G}_{n+1})|\mathcal{G}_n\big) = \mathrm{E}(Z|\mathcal{G}_n) = X_n.$$

Later we shall see that for any martingale $\{X_n\}$ with an added condition, one can identify a pair $(Z, \mathcal{G}_n)$ so that it becomes a Doob's martingale, $X_n = \mathrm{E}(Z|\mathcal{G}_n)$ almost surely. ▲

Example 4.1.5 (*Bienaymé–Galton–Watson branching process*) Suppose we have iid random variables $\{\xi_{n,k} : n \geq 1, k \geq 1\}$ which take values in $\{0, 1, 2, \ldots\}$, and have a finite mean m. Let $Z_0 := 1$ and for $n \geq 1$,

$$Z_n := \begin{cases} \xi_{n,1} + \xi_{n,2} + \cdots + \xi_{n,Z_{n-1}} & \text{if } Z_{n-1} > 0, \\ 0 & \text{if } Z_{n-1} = 0. \end{cases}$$

The idea is that we start with a single parent Z_0, called the root. The kth parent in the $(n-1)$th generation produces $\xi_{n,k}$ offspring. At the nth generation, the total number of progeny is Z_n.

Let $\mathcal{G}_0 := \{\emptyset, \Omega\}$ and $\mathcal{G}_n := \sigma(\xi_{j,k} : j \leq n, 1 \leq k < \infty)$. Then $(m^{-n}Z_n, \mathcal{G}_n)_{n \geq 0}$ is a martingale. This is easy to see by the following simple calculation.

$$\mathrm{E}(m^{-(n+1)}Z_{n+1}|\mathcal{G}_n) = m^{-(n+1)}\mathrm{E}\big(\mathbf{1}_{\{Z_n>0\}}\big(\xi_{n+1,1} + \cdots + \xi_{n+1,Z_n}\big)|\mathcal{G}_n\big)$$

$$= m^{-(n+1)}\mathbf{1}_{\{Z_n>0\}}\mathrm{E}\left(\sum_{k=1}^{Z_n}\xi_{n+1,k}|\mathcal{G}_n\right)$$

$$= m^{-(n+1)}\sum_{\ell=1}^{\infty}\mathbf{1}_{\{Z_n=\ell\}}\ell m$$

$$= m^{-n}Z_n = X_n.$$

▲

Urn models have a lot of applications, in particular in clinical trials. Analyses of the probabilistic properties of such models use of martingale theory. In the following example we introduce the simplest urn model and identify a relevant martingale structure.

Example 4.1.6 (*Pólya's Urn*) An urn has $b_0 > 0$ black, and $w_0 > 0$ white balls. A ball is drawn at random and returned to the urn with an additional new ball of the same color. This process is repeated. What happens to the proportions of the black and white balls if we continue the process? This scheme can be described by the following probabilistic construction. Let $\{U_n\}_{n\geq 1}$ be iid Uniform[0, 1] random variables. Let $B_0 := b_0 > 0$ and $W_0 := w_0 > 0$ be given. Define

$$\xi_n := \mathbf{1}_{\{U_n \leq \frac{B_{n-1}}{B_{n-1}+W_{n-1}}\}},$$
$$B_n := B_{n-1} + \xi_n,$$
$$W_n := W_{n-1} + (1 - \xi_n).$$

So, $\xi_n = 1$ indicates that the nth draw is black. Then B_n and W_n are the number of black and white balls in the urn before the $(n+1)$th draw. Clearly, $B_n + W_n = b_0 + w_0 + n$ (as one ball is added after each draw). Let $\mathcal{G}_n := \sigma(U_1, \ldots, U_n)$, so that ξ_n, B_n and W_n are all $\mathcal{G}_n$-measurable. Let

$$X_n := \frac{B_n}{B_n + W_n} = \frac{B_n}{b_0 + w_0 + n}$$

be the fraction of black balls at stage n. It is $\mathcal{G}_n$-measurable.

Observe that

$$\begin{aligned}
\mathrm{E}(B_n|\mathcal{G}_{n-1}) &= B_{n-1} + \mathrm{E}(\mathbf{1}_{\{U_n \leq X_{n-1}\}}|\mathcal{G}_{n-1}) \\
&= B_{n-1} + X_{n-1} \\
&= \frac{b_0 + w_0 + n}{b_0 + w_0 + n - 1} B_{n-1}.
\end{aligned}$$

Hence we have

$$\begin{aligned}
\mathrm{E}(X_n|\mathcal{G}_{n-1}) &= \frac{1}{b_0 + w_0 + n} \mathrm{E}(B_n|\mathcal{G}_{n-1}) \\
&= \frac{1}{b_0 + w_0 + n - 1} B_{n-1} = X_{n-1}.
\end{aligned}$$

Hence $(X_n, \mathcal{G}_n)$ is a martingale. ▲

For our next example of a martingale, we need the concepts of a Markov chain and a simple random walk on a graph.

Definition 4.1.3 (*Markov chain*) Random variables $\{X_n\}$, taking values in $S = \{x_i, i \geq 1\}$, is said to be a **Markov chain** with **state space** S if, for all $n \geq 2$ and $\{y_1, \ldots, y_n\} \subset S$,

$$P(X_n = y_n|X_1 = y_1, \ldots, X_{n-1} = y_{n-1}) = P(X_n = y_n|X_{n-1} = y_{n-1}). \quad (4.1)$$

The probabilities $p_{ij} = P(X_n = j|X_{n-1} = i)$, $i, j \in S$ are known as the **transition probabilities**. Equation (4.1) is known as the **Markov property**. ◇

Exercise 4.1.7 Verify that the simple random walk is a Markov chain, and identify its state space and transition probabilities.

We now define the simple random walk on a graph.

Definition 4.1.4 Suppose $G = (V, E)$ is a graph with a countable vertex set V, and edge set E. For any $i \in V$, let $\deg(i) < \infty$ be its degree. We write $j \sim i$ if there is at least one edge from i to j. Let $\{S_n\}$ be a sequence of V-valued random variables such that $S_0 = v$ where $v \in V$ is some designated vertex, and for $n \geq 2$,

$$P(S_n = j | S_1 = y_1, \ldots, S_{n-1} = i) = p_{ij} = \begin{cases} \frac{1}{\deg(i)} & \text{if } j \sim i, \\ 0 & \text{otherwise}. \end{cases}$$

Then S_n is a Markov chain and is a **simple random walk on the graph** G. ◇

Example 4.1.7 (*Discrete harmonic function*) Let $\{S_n\}_{n \geq 0}$ be a simple random walk on a graph G with a countable vertex set V, where each vertex has finite degree. Let $\varphi : V \to \mathbb{R}$ be a harmonic function, that is,

$$\varphi(i) = \frac{1}{\deg(i)} \sum_{j:j \sim i} \varphi(j) \quad \text{for all } i \in V.$$

Then $(\varphi(S_n), \mathcal{G}_n = \sigma(S_0, S_1, \ldots, S_n))_{n \geq 0}$ is a martingale. This can be seen as follows:

$$E(\varphi(S_{n+1})|\mathcal{G}_n) = \frac{1}{\deg(S_n)} \sum_{j:j \sim S_n} \varphi(j) \quad \text{(using the Markov property)}$$

$$= \varphi(S_n).$$

▲

4.2 Stopping Time

Let us go back to our game analogy. A player may try to devise strategies to improve his winnings. For example, based on the current information, he may decide to stop playing the game at that point or, may decide to play a number of additional times (which may be random). This notion is captured by the idea of a stopping time or a stop time given below. For technical reasons, we will allow a stopping time to take the value ∞.

Let $\{\mathcal{G}_n\}_{0 \leq n < \infty}$ be a filtration. We augment it by some σ-field $\mathcal{G}_\infty$, so that $\{\mathcal{G}_n\}_{0 \leq n \leq \infty}$ remains a filtration. Typically $\mathcal{G}_\infty = \bigvee_{0 \leq n < \infty} \mathcal{G}_n$.

Definition 4.2.1 (*Stopping time*) $\tau : \Omega \to \mathbb{N} \cup \{0, \infty\}$ is said to be a **stopping time** or a **stop time** with respect to filtration $\{\mathcal{G}_n\}_{0 \le n \le \infty}$ if,

$$\{\tau = n\} \in \mathcal{G}_n \quad \text{for all } 0 \le n \le \infty.$$

We will say τ is a $\{\mathcal{G}_n\}$-stop time, and will not identify the filtration if, either it is clear from the context, or it is not important to identify it. If the process starts at $n = 1$, then the minimum value of τ is 1. ◇

So, the decision to stop at time n is based only on the information $\mathcal{G}_n$. In the tradition of equating quantities that differ only on a null set, if $(\Omega, \mathcal{A}, P)$ is a probability space, τ_1 is a stopping time, and τ_2 is another random variable which is almost surely equal to τ_1, then τ_2 is also considered to be a stop time (with respect to the same filtration). Technically, it suffices to enlarge the σ-fields in the filtration by taking their completions.

Exercise 4.2.1 Suppose $(\Omega, \mathcal{A}, P)$ is a probability space. Suppose τ is a random variable taking values in $\mathbb{N} \cup \{0, \infty\}$. Show that τ is a $\{\mathcal{G}_n\}$-stop time if and only if

$$\{\omega : \tau(\omega) \le k\} \in \mathcal{G}_k \quad \text{for all } k \ge 0.$$

A stop time τ is called **finite** if $P\{\tau < \infty\} = 1$. It is called **bounded** if, for some finite constant K, $P\{\tau \le K\} = 1$. In these cases, we do not need $\mathcal{G}_\infty$.

If τ is a finite stop time, then for any sequence of random variables $\{X_n\}_{0 \le n < \infty}$, $X_\tau : \Omega \to \mathbb{R}$ is defined by $X_\tau(\omega) := X_n$ on the set $\{\tau = n\}$, $0 \le n < \infty$. It is easy to check that X_τ is a random variable.

If $P\{\tau = \infty\} > 0$, and there is a random variable X_∞, then X_τ is well-defined and continues to be a random variable.

Example 4.2.1 Check that in the following, the first two yield stop times but the third one does not.

(a) Let $(X_n, \mathcal{G}_n)_{n \ge 0}$ be an adapted sequence, and let A be an appropriate Borel set. Define

$$\tau_A := \inf\{n : X_n \in A\}.$$

Then τ_A is said to be the (first) **hitting time** of A.

(b) Suppose $\{\xi_j\}_{j \ge 1}$ are iid random variables taking values 1 and 0. Let $\mathcal{G}_n := \sigma(\xi_1, \ldots, \xi_n)$. Define

$$\tau := \inf\{n \ge 4 : (\xi_{n-3}, \xi_{n-2}, \xi_{n-1}, \xi_n) = (1, 0, 1, 1)\}.$$

If we identify the values 1 and 0 of ξ_n with head-tail outcomes for coin tosses, then τ is the first time that we see the string HTHH.

(c) Similar to the first hitting time, let $N \le \infty$ and

$$T = \sup\{n \le N : X_n \notin A\}.$$

4.2 Stopping Time

Thus T gives the last visit time to the set A before time N. ▲

We list some basic properties of stop times that can be verified easily.

Lemma 4.2.1 *(i) If τ is deterministic or constant almost surely, then it is a stop time. (ii) If τ_1 and τ_2 are stopping times, then the variables $\tau_1 \vee \tau_2$, $\tau_1 \wedge \tau_2$ and $\tau_1 + \tau_2$ are also stopping times.* ◆

Proof Part (i) is trivial. To prove Part (ii), observe that

$$\{\tau_1 \vee \tau_2 \leq n\} = \{\tau_1 \leq n\} \cap \{\tau_2 \leq n\},$$
$$\{\tau_1 \wedge \tau_2 \leq n\} = \{\tau_1 \leq n\} \cup \{\tau_2 \leq n\}, \text{ and}$$
$$\{\tau_1 + \tau_2 \leq n\} = \cup_{k=0}^{n} \{\tau_1 = k\} \cap \{\tau_2 \leq n - k\}.$$

Now note that all events on the right side are $\mathcal{G}_n$-measurable. □

Exercise 4.2.2 Let τ_1 and τ_2 be stop times. Is $\tau_1 \tau_2$ a stop time? Suppose $P\{\tau_1 \geq \tau_2\} = 1$. Is $\tau_1 - \tau_2$ a stop time?

Note that for any stop time τ, $\tau \wedge n$ is always a bounded stop time. Then we can define a **stopped process** as follows.

Definition 4.2.2 (*Stopped process*) Let $(X_n, \mathcal{G}_n)_{n \geq 0}$ be adapted, and τ be a $\{\mathcal{G}_n\}$-stop time. Then the process stopped at τ is defined as

$$X_n^\tau := X_{\tau \wedge n}.$$

It is also referred to as a **stopped process**. ◇

Lemma 4.2.2 (*Stopped process is adapted*) *If $(X_n, \mathcal{G}_n)_{n \geq 0}$ is an adapted sequence and τ is a $\{\mathcal{G}_n\}$-stop time, then $(X_n^\tau, \mathcal{G}_n)_{n \geq 0}$ is also an adapted sequence.* ◆

Proof The proof is immediate from the following decomposition into $\mathcal{G}_n$-measurable events.

$$\{X_n^\tau \in A\} = \{X_n^\tau \in A, \tau \leq n\} \cup \{X_n^\tau \in A, \tau > n\}$$
$$= \cup_{k=1}^n \left(\{X_k \in A\} \cap \{\tau = k\}\right) \cup \left(\{X_n \in A\} \cap \{\tau > n\}\right).$$

To finish the proof, note that each set on the right side belongs to $\mathcal{G}_n$. □

Definition 4.2.3 (*Stopped σ-field*) For a $\{\mathcal{G}_n\}$-stop time τ,

$$\mathcal{G}_\tau := \{A \in \mathcal{A} : A \cap \{\tau \leq k\} \in \mathcal{G}_k \text{ for all } k \geq 0\}$$

is called the **stopped σ-field**. ◇

Exercise 4.2.3 (a) Check that a stopped σ-field is indeed a σ-field.

(b) If τ is a constant k, show that $\mathcal{G}_\tau = \mathcal{G}_k$.

(c) Let τ_1, τ_2 be two stop times, and $\tau_1 \leq \tau_2$. Show that $\mathcal{G}_{\tau_1} \subset \mathcal{G}_{\tau_2}$.

(d) If $(X_n, \mathcal{G}_n)$ is adapted, and τ is a finite stop time, then show that X_τ is $\mathcal{G}_\tau$-measurable.

Exercise 4.2.4 Let $(X_n, \mathcal{G}_n)_{n \geq 1}$ be an adapted sequence where $\{X_n\}$ is integrable. Show that it is a martingale if and only if $\mathrm{E}(X_k | \mathcal{G}_\tau) = X_\tau$ for all $\{P\mathcal{G}_n\}$-stop times $\tau \leq k, k \geq 1$.

4.3 Doob's Optional Stopping Theorem

Exercise 4.3.1 Let $(X_n, \mathcal{G}_n)_{n \geq 0}$ be an adapted sequence where each X_n is integrable. If τ is a $\{\mathcal{G}_n\}$-stop time, then show that for every n, X_n^τ is also integrable.

Theorem 4.3.1 *Let $(X_n, \mathcal{G}_n)_{n \geq 0}$ be an adapted sequence and τ be a $\{\mathcal{G}_n\}$-stop time. Then $(X_n^\tau, \mathcal{G}_n)_{n \geq 0}$ is a martingale, a super-martingale or a sub-martingale according as $(X_n, \mathcal{G}_n)_{n \geq 0}$ is a martingale, a super-martingale, or a sub-martingale.* ◆

Proof We have already seen that $(X_n^\tau, \mathcal{G}_n)$ is adapted and X_n^τ is integrable for each n. Note that for every n, we can write X_n^τ as

$$X_n^\tau = \sum_{k=0}^{n-1} X_k \mathbf{1}_{\{\tau = k\}} + X_n \mathbf{1}_{\{\tau > n-1\}}.$$

First suppose that $(X_n, \mathcal{G}_n)_{n \geq 0}$ is a martingale. Then

$$\mathrm{E}(X_{n+1}^\tau | \mathcal{G}_n) = \mathrm{E}\left(\sum_{k=0}^n X_k \mathbf{1}_{\{\tau=k\}} + X_{n+1} \mathbf{1}_{\{\tau > n\}} | \mathcal{G}_n\right)$$

$$= \sum_{k=0}^n X_k \mathbf{1}_{\{\tau=k\}} + \mathrm{E}(X_{n+1} | \mathcal{G}_n) \mathbf{1}_{\{\tau > n\}} \quad \text{(by Exercise 4.2.1)}$$

$$= \sum_{k=0}^{n-1} X_k \mathbf{1}_{\{\tau=k\}} + X_n \mathbf{1}_{\{\tau \geq n\}} = X_n^\tau.$$

Thus $(X_n^\tau, \mathcal{G}_n)$ is a martingale. Proofs of the other claims are similar and are omitted. □

Remark 4.3.1 A consequence of the martingale property above is that

$$\mathrm{E}(X_{\tau \wedge n}) = \mathrm{E}(X_n^\tau) = \mathrm{E}(X_0^\tau) = \mathrm{E}(X_0).$$

4.3 Doob's Optional Stopping Theorem

On the other hand, if τ is finite, then $X_{\tau \wedge n} \to X_\tau$ a.s. as $n \to \infty$. Does it follow that

$$E(X_\tau) \stackrel{?}{=} E(X_0) \quad \text{or equivalently,} \quad E(\lim_{n \to \infty} X_{\tau \wedge n}) \stackrel{?}{=} \lim_{n \to \infty} E(X_{\tau \wedge n}). \tag{4.2}$$

The answer is no in general. ●

Example 4.3.1 Suppose $\{\xi_k\}_{k \geq 1}$ are iid with $P(\xi_1 = \pm 1) = 1/2$. Define $X_0 := 0$, $X_n := \sum_{k=1}^n \xi_k$, $\mathcal{G}_n := \sigma(\xi_1, \ldots, \xi_n)$ and $\mathcal{G}_0 = \{\emptyset, \Omega\}$. We know that $(X_n, \mathcal{G}_n)_{n \geq 0}$ is a martingale. Define τ as $\tau := \inf\{n : X_n = +1\}$. Clearly τ is a stop time, $P(\tau < \infty) = 1$, and $X_\tau \equiv 1$ a.s. Hence $E(X_\tau) = 1 \neq 0 = E(X_0)$. ▲

We now present a sufficient condition for equality to hold in (4.2).

Theorem 4.3.2 (Doob's Optional stopping theorem (OST)) *Suppose $(X_n, \mathcal{G}_n)$ is a sub-martingale and τ_1, τ_2 are stopping times where $P(\tau_1 \leq \tau_2 \leq n) = 1$ for some $n \in \mathbb{N}$. Then, $(X_{\tau_i}, \mathcal{G}_{\tau_i})_{i=1,2}$ is a sub-martingale.* ◆

Proof Integrability of X_{τ_1} and X_{τ_2} can be quickly verified by using Exercise 4.3.1 and we omit the details. Consider first the sub-martingale case. It suffices to show that

$$\int_A (X_{\tau_2} - X_{\tau_1}) dP \geq 0 \quad \text{for all} \quad A \in \mathcal{G}_{\tau_1}.$$

Fix $A \in \mathcal{G}_{\tau_1}$. Note that for all $k \in \mathbb{N}$,

$$A \cap \{\tau_1 < k \leq \tau_2\} = A \cap \{\tau_1 \leq k-1\} \cap \{\tau_2 \leq k-1\}^c \in \mathcal{G}_{k-1}.$$

Therefore,

$$\int_A (X_{\tau_2} - X_{\tau_1}) dP = \int_A \sum_{k=\tau_1+1}^{\tau_2} (X_k - X_{k-1}) dP$$

$$= \sum_{k=2}^n \int_{A \cap \{\tau_1 < k \leq \tau_2\}} (X_k - X_{k-1}) dP$$

$$= \sum_{k=2}^n \int_{A \cap \{\tau_1 < k \leq \tau_2\}} (E(X_k | \mathcal{G}_{k-1}) - X_{k-1}) dP$$

$$\geq 0 \quad \text{(by the sub-martingale property)}.$$

□

It is quite obvious that results analogous to Theorem 4.3.2 hold for super-martingales and martingales.

Exercise 4.3.2 Show that Theorem 4.3.1 follows from Theorem 4.3.2.

Several consequences of Theorem 4.3.2 are given below.

Corollary 4.3.1 (Condition for $E(X_\tau) = E(X_0)$) *Let $(X_n, \mathcal{G}_n)_{n\geq 0}$ be a martingale. Suppose that τ is a finite $\mathcal{G}_n$-stop time. Then, X_τ is integrable and*

$$E(X_\tau) = E(X_0)$$

under any of the following conditions:

(i) The sequence $\{X_{\tau \wedge n} : n \geq 1\}$ is UI.
(ii) For some constant $K < \infty$, $P(\sup_n |X_{\tau \wedge n}| \leq K) = 1$.
(iii) $E(\tau) < \infty$, and $\sup_n E(|X_{n+1} - X_n||\mathcal{G}_n) < K$ a.s. for some constant $K < \infty$. ◆

Proof For every n, $\tau \wedge n$ is bounded. Hence by Doob's OST Theorem 4.3.2, the pair $(X_1, \mathcal{G}_1)$, $(X_{\tau \wedge n}, \mathcal{G}_{\tau \wedge n})$ is a martingale. Thus

$$E(X_1) = E(X_{\tau \wedge n}) \text{ for all } n \geq 1. \tag{4.3}$$

On the other hand, since τ is finite,

$$X_{\tau \wedge n} \to X_\tau \text{ almost surely, as } n \to \infty. \tag{4.4}$$

(i) By (4.4), UI of $\{X_{\tau \wedge n} : n \geq 1\}$ and Theorem 3.4.1,

$$\lim_{n \to \infty} E(X_{\tau \wedge n}) = E(X_\tau).$$

This, in conjunction with (4.3), completes the proof of Part (i).

(ii) The arguments are similar, except that DCT is used in place of UI.

(iii) To prove Part (iii) note that

$$X_{\tau \wedge n} - X_0 = \sum_{k=0}^{n-1} (X_{k+1} - X_k) \mathbf{1}_{\{\tau > k\}}.$$

We now show that

$$Y = \sum_{k=0}^{\infty} |X_{k+1} - X_k| \mathbf{1}_{\{\tau > k\}} < \infty, \text{ and } E(Y) < \infty. \tag{4.5}$$

Observe that

$$E(Y) = \sum_{k=0}^{\infty} E\left(|X_{k+1} - X_k| \mathbf{1}_{\{\tau > k\}}\right)$$

$$= \sum_{k=0}^{\infty} E\left(E(|X_{k+1} - X_k| \mathbf{1}_{\{\tau > k\}} | \mathcal{G}_k)\right)$$

4.3 Doob's Optional Stopping Theorem

$$= \sum_{k=0}^{\infty} E\left(1_{\{\tau>k\}} E(|X_{k+1} - X_k| \,|\, \mathcal{G}_k)\right)$$

$$\leq K \sum_{k=0}^{n-1} P(\tau > k)$$

$$= K E(\tau) < \infty.$$

This establishes (4.5). Since $|X_{\tau \wedge n} - X_0| \leq Y$ for all n, and $E(Y) < \infty$, we can apply DCT to (4.4) to conclude the proof of Part (iii). □

Exercise 4.3.3 Suppose $(X_n, \mathcal{G}_n)_{n \geq 0}$ is an adapted sequence. Show that if $\{X_n\}$ is UI then $\{X_{\tau \wedge n}\}$ is also UI for any finite $\mathcal{G}_n$-stop time τ.

Exercise 4.3.4 Suppose $\{S_n : n \geq 0\}$ is a simple symmetric random walk. Let

$$\tau = \inf\{n \geq 0 : S_n = 1\}.$$

Show that $\tau < \infty$ almost surely, and $E(\tau) = \infty$.

Example 4.3.2 (*Ruin probability*) Let $-a < 0 < b$ be integers. A gambler starts with a capital of a units and plays a series of games. Her earnings for the games are iid ξ_n, $P(\xi_n = 1) = P(\xi_n = -1) = 1/2$. Let S_n be her cumulative earnings at time n (not counting the initial capital a). She will quit the first time S_n reaches b or $-a$ (ruin). What is the probability that she reaches $-a$ before b? This **ruin probability** is given by

$$p_r := P\left(S_n = -a \text{ for some } n \geq 1, \text{ and } S_m \neq b \text{ for all } 1 \leq m \leq n-1\right).$$

To compute this probability, let $\mathcal{G}_n := \sigma(S_1, \ldots, S_n)$. Define the $\{\mathcal{G}_n\}$-stopping time τ as

$$\tau := \inf\{n \in \mathbb{N} : S_n \in \{-a, b\}\}.$$

Observe that $(S_n, \mathcal{G}_n)_{n \geq 0}$ is a martingale and $|S_{\tau \wedge n}| \leq a + b$. Hence, by Corollary 4.3.1,

$$E(S_\tau) = E(S_1) = 0.$$

On the other hand, S_τ takes values $-a$ and b with probabilities p_r and $1 - p_r$ respectively.
Therefore,

$$0 = E(S_\tau) = -ap_r + b(1 - p_r).$$

Upon solving,

$$p_r = \frac{b}{a+b}.$$

We can also calculate $E(\tau)$. For this, we use the zero mean martingale $(S_n^2 - n, \mathcal{G}_n)$.

Note that $E(\tau) \leq E(S_\tau^2) < \infty$. Hence we can apply Part (iii) of Corollary 4.3.1 to obtain
$$E(S_\tau^2 - \tau) = E(S_1^2 - 1) = 0.$$

This gives
$$E(\tau) = E(S_\tau^2) = a^2 p_r + b^2(1 - p_r)$$
$$= a^2 \frac{b}{a+b} + b^2 \frac{a}{a+b} = ab.$$

▲

Example 4.3.3 (*Non-symmetric random walk*) Let us go back to Example 4.3.2. So, let $-a < 0 < b$ be two integers. But now $\{\xi_n\}$ are iid with $P(\xi_n = 1) = p$, $P(\xi_n = -1) = 1 - p$ and $p > q > 0$. Let $S_n := \sum_{i=1}^n \xi_i$ and $\mathcal{G}_n$ be as before. We wish to calculate the probability of ruin in finite time. Define three stop times
$$\tau_{-a} := \inf\{n \in \mathbb{N} : S_n = -a\}, \quad \tau_b := \inf\{n \in \mathbb{N} : S_n = b\}, \quad \tau := \tau_{-a} \wedge \tau_b.$$

Then the probability of ruin in finite time equals $P(\tau_{-a} < \infty)$.

To compute this quantity, we now use a different martingale. Let $M_n := (p^{-1}q)^{S_n}$. We show that $(M_n, \mathcal{G}_n)$ is a martingale. Clearly it is an adapted sequence. Further, since $|M_n| \leq (p/q)^n$, it is integrable.

Now,
$$E(M_{n+1}|\mathcal{G}_n) = (p^{-1}q)^{S_n} E\big((p^{-1}q)^{\xi_{n+1}}|\mathcal{G}_n\big)$$
$$= (p^{-1}q)^{S_n} E\big((p^{-1}q)^{\xi_{n+1}}\big)$$
$$= (p^{-1}q)^{S_n} = M_n.$$

Hence $(M_n, \mathcal{G}_n)$ is a martingale with $E(M_n) = 1$ for all n. We leave it to the reader to check that $P(\tau < \infty) = 1$.

Since $M_{\tau \wedge n}$ is bounded between $(p^{-1}q)^{-a}$ and $(p^{-1}q)^b$, we can apply Doob's OST Theorem 4.3.2 to get $E(M_\tau) = 1$. This implies that
$$1 = (p^{-1}q)^{-a} P(\tau = \tau_{-a}) + (p^{-1}q)^b P(\tau = \tau_b).$$

Of course, $P(\tau = \tau_{-a}) + P(\tau = \tau_b) = 1$. Using these two equations, it follows that
$$P(\tau_{-a} < \tau_b) = \frac{(p^{-1}q)^b - 1}{(p^{-1}q)^b - (p^{-1}q)^{-a}}. \tag{4.6}$$

Since it takes at least b plays to reach b from 0, $\tau_b \geq b$. Thus as $b \to \infty$, $\tau_b \to \infty$, and

4.3 Doob's Optional Stopping Theorem

$$\mathbf{1}_{\{\tau=\tau_{-a}\}} = \mathbf{1}_{\{\tau_{-a}<\tau_b\}} \to \mathbf{1}_{\{\tau_{-a}<\infty\}}.$$

Letting $b \to \infty$, in (4.6), and using the fact that $(q/p)^b \to 0$,

$$P(\tau_{-a} < \infty) = (p^{-1}q)^a.$$

We can also get information on other quantities of interest. For example, let us show that min S_n is integrable.

Since $\{\tau_{-x} < \infty\} = \{\min S_n < -x\}$ for any $x > 0$, we have

$$\mathrm{E}(\min S_n) = -\sum_{x=0}^{\infty} x P(\min S_n = -x)$$

$$= -\sum_{x=1}^{\infty} x\left((p^{-1}q)^{x-1} - (p^{-1}q)^x\right)$$

$$= -(pq^{-1} - 1)\sum_{x=1}^{\infty} x\,(p^{-1}q)^x > -\infty.$$

We leave it to the reader to check that $\mathrm{E}(\tau_b) = (p-q)^{-1}b$. ▲

We end this section with a "converse" of the OST Theorem 4.3.2.

Theorem 4.3.3 (Converse of OST) *Let $(X_n, \mathcal{G}_n)_{n\geq 0}$ be an adapted sequence. Suppose for all bounded stopping times τ, $\mathrm{E}(X_\tau) = \mathrm{E}(X_0)$. Then $(X_n, \mathcal{G}_n)_{n\geq 0}$ is a martingale.* ◆

Proof Fix $n \geq 1$ and let $G = \{\mathrm{E}(X_{n+1}|\mathcal{G}_n) > X_n\}$. Note that $G \in \mathcal{G}_n$. Define the bounded stopping time τ as $\tau := (n+1)\mathbf{1}_G + n\mathbf{1}_{G^c}$. Then $X_\tau = X_{n+1}\mathbf{1}_G + X_n\mathbf{1}_{G^c}$. Thus

$$Z := \mathrm{E}(X_\tau|\mathcal{G}_n) = \mathrm{E}(X_{n+1}|\mathcal{G}_n)\mathbf{1}_G + X_n\mathbf{1}_{G^c}.$$

Then we have $Z \geq X_n$ and the inequality is strict on G. Also we have

$$\mathrm{E}(Z) = \mathrm{E}(X_\tau) = \mathrm{E}(X_0) = \mathrm{E}(X_n).$$

The last equality holds since n can also be considered as a bounded stopping time. This implies $P(G) = 0$ and hence $\mathrm{E}(X_{n+1}|\mathcal{G}_n) \leq X_n$ almost surely. The opposite inequality follows by a similar argument, and this shows that $(X_n, \mathcal{G}_n)_{n\geq 0}$ is a martingale. □

Exercise 4.3.5 Suppose $(X_n, \mathcal{G}_n)$ is an adapted sequence, Show that it is a submartingale if and only if for all bounded $\mathcal{G}_n$-stopping times $\tau \geq \sigma \geq 0$ a.s., $\mathrm{E}(X_\tau) \geq \mathrm{E}(X_\sigma)$.

4.4 Wald's Identities

An important consequence of Doob's optional stopping theorem are the identities of Wald (1944, 1945) developed in the context of sequential analysis. They involve partial sums of independent random variables, and we will consider them to be identically distributed.

Theorem 4.4.1 (Wald's first identity) *Let $\{X_i\}_{i\geq 1}$ be a sequence of iid real-valued r.vs. Let $S_n := \sum_{i=1}^{n} X_i$, $\mathcal{G}_n := \sigma(X_1, \ldots, X_n)$. Let τ be a $\mathcal{G}_n$-stopping time. If $\mathrm{E}(|X_1|) < \infty$ and $\mathrm{E}(\tau) < \infty$, then $\mathrm{E}(|S_\tau|) < \infty$ and*

$$\mathrm{E}(S_\tau) = \mathrm{E}(\tau)\mathrm{E}(X_1). \tag{4.7}$$

♦

Proof First assume X_i are non-negative. For each n, $\tau \wedge n$ is a bounded stop time, and $(S_n - n\mathrm{E}(X_1), \mathcal{G}_n)$ is a martingale.
So, by Theorem 4.3.2,

$$\mathrm{E}(S_{\tau \wedge n}) = \mathrm{E}(\tau \wedge n)\mathrm{E}(X_1). \tag{4.8}$$

Since $\tau \wedge n \uparrow \tau$ and $S_{\tau \wedge n} \uparrow S_\tau$, by MCT, applied to (4.8), we get (4.7). Now, $\mathrm{E}(\tau) < \infty$ implies that $\mathrm{E}(S_\tau) < \infty$. This proves the result for non-negative random variables.

Now for the general case $\mathrm{E}(|X_1|) < \infty$, the above arguments imply that $\mathrm{E}(\sum_{k=1}^{\tau} |X_k|) < \infty$. Also, $|S_{\tau \wedge n}| \leq \sum_{k=1}^{\tau} |X_k|$ and $S_{\tau \wedge n} \to S_\tau$.
Hence by DCT we have

$$\mathrm{E}(S_\tau) = \lim_{n \to \infty} \mathrm{E}(S_{\tau \wedge n}) = \lim_{n \to \infty} \mathrm{E}(\tau \wedge n)\mathrm{E}(X_1).$$

By MCT it follows that $\mathrm{E}(\tau \wedge n) \uparrow \mathrm{E}(\tau)$, which completes the proof. □

Theorem 4.4.2 (Wald's second identity) *Let $\{X_i\}$, $\{\mathcal{G}_n\}$, τ be as in Theorem 4.4.1. Suppose $\mathrm{E}(X_1) = 0$ and $\sigma^2 = \mathrm{E}(X_1^2) < \infty$. Then*

$$\mathrm{E}(S_\tau^2) = \sigma^2 \mathrm{E}(\tau).$$

♦

Proof Consider the martingale $(M_n := S_n^2 - n\sigma^2, \mathcal{G}_n)_{n \geq 1}$. By Theorem 4.3.1 or optional stopping Theorem 4.3.2, we have

$$\mathrm{E}(S_{\tau \wedge n}^2) = \mathrm{E}(\tau \wedge n)\sigma^2. \tag{4.9}$$

The right side converges to $\mathrm{E}(\tau)\sigma^2$ by MCT.
By Fatou's lemma,

4.4 Wald's Identities

$$E(S_\tau^2) \le \liminf_{n\to\infty} E(S_{\tau\wedge n}^2) = E(\tau)\sigma^2 < \infty.$$

Finite variance of $\{X_i\}$ implies $E(X_i X_j)$ is finite. Note that for each i, X_i is independent of $\mathcal{G}_{i-1}$ and $\mathbf{1}_{\{\tau \ge i\}} \in \mathcal{G}_{i-1}$. Hence we have

$$E(X_i \mathbf{1}_{\{\tau \ge i\}} X_j \mathbf{1}_{\{\tau \ge j\}}) = 0 \text{ for } i \ne j.$$

This implies that

$$\sup_{m \ge n} E((S_{\tau\wedge m} - S_{\tau\wedge n})^2) = \sup_{m\ge n} \sum_{i=n+1}^m E(X_i \mathbf{1}_{\{\tau \ge i\}})^2$$
$$\le \sup_{m \ge n} \sum_{i > n} E(X_i^2 \mathbf{1}_{\{\tau \ge i\}})$$
$$\le \sum_{i>n} E(X_i^2) E(\mathbf{1}_{\{\tau \ge i\}})$$
$$= E(X_1^2) \sum_{i>n} E(\mathbf{1}_{\{\tau \ge i\}}) \to 0 \text{ as } n \to \infty,$$

since $E(\tau) < \infty$.

Thus $\{S_{\tau\wedge n}\}$ is Cauchy in L^2, and hence converges in L^2. But then $E(S_{\tau\wedge n}^2) \to E(S_\tau^2)$. Now, the use of (4.9) completes the proof. □

Theorem 4.4.3 (Wald's third identity) *Let $\{X_n\}$ and $\{\mathcal{G}_n\}$ be as in Theorem 4.4.1. Assume that $E(\exp(\theta X_1)) = \exp(\psi(\theta)) < \infty$ for all $\theta > 0$, and for some function ψ. Then for every bounded stopping time τ, $E(\exp(\theta S_\tau - \tau\psi(\theta))) = 1$.* ◆

Proof Since τ is bounded, the result follows directly from OST using the nonnegative martingale $M_n = \exp(\theta S_n - n\psi(\theta))$. □

Example 4.4.1 Let $\{S_n\}_{n \ge 0}$ be a simple symmetric random walk with $S_0 = 0$. Let

$$\tau := \begin{cases} \inf\{n : S_n = 1\} & \text{if there is such a finite } n, \\ \infty & \text{if there is no such finite } n. \end{cases}$$

Observe that from Exercise 4.3.4 it follows that $P(\tau < \infty) = 1$. Let us derive the moment generating function of τ using Wald's third identity. Note that for $\theta > 0$,

$$E(e^{\theta X_1}) = \frac{e^\theta + e^{-\theta}}{2} = \cosh(\theta).$$

Then by Theorem 4.4.3 we have

$$E\left(\cosh(\theta)^{-(\tau\wedge n)} e^{\theta S_{\tau\wedge n}}\right) = 1.$$

As $n \to \infty$, $\cosh(\theta)^{-(\tau \wedge n)} e^{\theta S_{\tau \wedge n}} \to \cosh(\theta)^{-\tau} e^{\theta S_\tau}$. Since $S_{\tau \wedge n} \leq n$, for $\theta > 0$, we have $e^{\theta S_{\tau \wedge n}} \leq e^\theta$. Also, $\cosh(\theta) > 1$ so $\cosh(\theta)^{-(\tau \wedge n)} \leq 1$. Hence by DCT,

$$\mathrm{E}\bigl(\cosh(\theta)^{-\tau} e^{\theta S_\tau}\bigr) = 1.$$

So it follows that
$$\mathrm{E}\bigl(\cosh(\theta)^{-\tau}\bigr) = e^{-\theta}. \tag{4.10}$$

Let $z = \frac{1}{\cosh(\theta)}$. Then it is easily seen that $e^{-\theta} = \frac{1-\sqrt{1-z^2}}{z}$. Hence, we can rewrite (4.10) as

$$\mathrm{E}(z^\tau) = \frac{1 - \sqrt{1-z^2}}{z}, \quad 0 < z < 1.$$

▲

4.5 Maximal Inequality

Probabilistic control on the maximum of a process is a very important tool in probability theory. There are numerous probability inequalities to gain this control. We shall state and prove some basic maximal inequalities, especially in connection to sub-martingales. These will be used later to derive many properties of martingales.

Due to its historic importance, we state and prove Kolmogorov's Maximal Inequality for sums of independent random variables, even though we will soon establish a much more general maximal inequality.

Theorem 4.5.1 (Kolmogorov's Maximal Inequality) *Let $\{X_i\}$ be independent random variables such that for all i, $\mathrm{E}(X_i) = 0$ and $\mathrm{E}(X_i^2) < \infty$. Let $S_n := X_1 + \cdots + X_n$, $n \geq 1$. Then*

$$P\bigl(\max_{1 \leq k \leq n} |S_k| \geq t\bigr) \leq \frac{1}{t^2} \mathrm{E}(S_n^2), \quad \text{for all } t > 0.$$

◆

Proof For $n \geq 1$, let $\mathcal{G}_n := \sigma(X_1, \ldots, X_n)$. Then clearly $(S_n, \mathcal{G}_n)$ is a martingale and hence $(S_n^2, \mathcal{G}_n)$ is a sub-martingale. Let

$$\tau := \min\{k : |S_k| \geq t\}.$$

Then for every n, $\tau \wedge n$ is a $\mathcal{G}_n$-stop time bounded by n. Hence the pair $(S_{\tau \wedge n}^2, \mathcal{G}_{\tau \wedge n})$, $(S_n^2, \mathcal{G}_n)$ is a sub-martingale. Therefore it immediately follows that $\mathrm{E}(S_{\tau \wedge n}^2) \leq \mathrm{E}(S_n^2)$.

Then we have

4.5 Maximal Inequality

$$P(\max_{k \leq n} |S_k| \geq t) = P(S_{\tau \wedge n}^2 \geq t^2)$$

$$\leq \frac{1}{t^2} E(S_{\tau \wedge n}^2) \text{ (by Markov inequality)}$$

$$\leq \frac{1}{t^2} E(S_n^2).$$

$\square$

Remark 4.5.1 Kolmogorov's Maximal Inequality is remarkable since the maximum of the entire path $S_1, \ldots, S_n$ is controlled in terms of the end-point S_n alone. There is nothing fallacious regarding the inequality $E(S_{\tau \wedge n}^2) \leq E(S_n^2)$— if the path goes beyond $(-t, t)$, then there is a significant probability for the value of S_n to also be large. ●

Exercise 4.5.1 Construct a direct proof of Theorem 4.5.1 avoiding sub-martingales.

Exercise 4.5.2 Suppose $\{X_i\}$ are iid with zero mean and unit variance. Recall that the Central Limit Theorem says that $S_n/\sqrt{n}$ converges in distribution to a standard Gaussian distribution. Using Theorem 4.5.1, show that $\{\max_{1 \leq k \leq n} |S_k|/\sqrt{n}\}$ is tight. That is, given $\varepsilon > 0$, there exists a K such that $P\{\max_{1 \leq k \leq n} |S_k|/\sqrt{n} > K\} < \varepsilon$ for all $n \geq 1$. It is much harder to prove that $\{\max_{1 \leq k \leq n} |S_k|/\sqrt{n}\}$ also converges in distribution.

Theorem 4.5.2 (Doob's Maximal Inequality) Suppose $(X_n, \mathcal{G}_n)$ is a sub-martingale. Then for every $t > 0$, and $n \geq 1$,

$$P\left(\max_{1 \leq i \leq n} X_i \geq t\right) \leq \frac{1}{t} E\left(X_n \mathbf{1}_{\{\max_{1 \leq i \leq n} X_i \geq t\}}\right) \leq \frac{1}{t} E(|X_n|).$$

♦

Proof Define

$$\tau := \min\{k : X_k \geq t\} \wedge n.$$

Note that τ is a stop time bounded by n. Hence X_τ is integrable. Therefore, by Doob's OST Theorem 4.3.2, the pair $(X_\tau, \mathcal{G}_\tau)$, $(X_n, \mathcal{G}_n)$ is a sub-martingale. Now define

$$M_k := \max_{1 \leq i \leq k} X_i, \ 1 \leq k \leq n.$$

We first show that $\{M_n \geq t\} \in \mathcal{G}_\tau$ for every n and t. This follows from

$$\{M_n \geq t\} \cap \{\tau \leq k\} = \{M_k \geq t\} \in \mathcal{G}_k \text{ for all } 1 \leq k \leq n-1,$$
$$\{M_n \geq t\} \cap \{\tau \leq k\} = \{M_n \geq t\} \in \mathcal{G}_n \subset \mathcal{G}_k \text{ for all } k \geq n.$$

Thus, $\{M_n \geq t\} \in \mathcal{G}_\tau$. Hence,

$$tP(M_n \geq t) \leq \int_{\{M_n \geq t\}} X_\tau \, dP \text{ (by definition of } M_n, \text{ and } \tau \leq n)$$
$$\leq \int_{\{M_n \geq t\}} X_n \, dP \text{ (sub-martingale property)}$$
$$\leq \int_{\{M_n \geq t\}} |X_n| \, dP$$
$$\leq E(|X_n|).$$

This completes the proof. □

The following form of Doob's Maximal Inequality (Theorem 4.5.2) is often used. Its proof is immediate once we observe that $(|X_n|, \mathcal{G}_n)$ is a sub-martingale.

Corollary 4.5.1 *Let $(X_n, \mathcal{G}_n)$ be a martingale. Then for $t > 0$, and $n \geq 1$,*

$$P\left(\max_{1 \leq i \leq n} |X_i| \geq t\right) \leq \frac{1}{t} E\left(|X_n| \mathbf{1}_{\{\max_{1 \leq i \leq n} |X_i| \geq t\}}\right) \leq \frac{1}{t} E(|X_n|).$$

◆

4.6 Abracadabra

We now delve into a "tricky exercise" stated in Sect. 4.9, page 45 of Williams (1991). Later in Exercise E10.6, page 233 it is asked why the answer given there is intuitively obvious from martingale theory. We will elevate this exercise to the status of a Theorem and provide a proof based on the lecture notes of Caravenna (2016).

Theorem 4.6.1 *Suppose a typist strikes the letter keys of the English alphabet, and with each strike she is equally likely to type any of the 26 letters, independent of the other strikes. Let τ be the number of strokes needed to produce the sequence* abracadabra. *Then*

$$E(\tau) = 26^{11} + 26^4 + 26. \tag{4.11}$$

◆

Proof Let $\{U_i\}_{i \geq 1}$ be independent "random variables", each uniformly distributed on the set $E = \{\text{a}, \text{b}, \text{c}, \text{d}, \ldots, \text{x}, \text{y}, \text{z}\}$. For $m \leq n$, define

$$U_{[m,n]} := (U_m, U_{m+1}, \ldots, U_n).$$

Let τ be the random time at which the word abracadabra first appears, that is,

$$\tau = \min\{n \geq 11 : U_{[n-10,n]} = \text{abracadabra}\}.$$

4.6 Abracadabra

Let $\mathcal{G}_n = \sigma(U_1, U_2, \ldots, U_n)$. We visualize a sequence of probabilistic games that gamblers play as follows. Let $x_1 = \text{a}$, $x_2 = \text{b}, \ldots, x_{11} = \text{a}$ denote the successive letters of abracadabra.

The first gambler enters the game with an initial capital of 1. She bets that $U_1 = x_1$. If she loses, her capital becomes zero, and she must stop playing. If she wins, her capital becomes 26, and she bets that $U_2 = x_2$. If she loses, her capital becomes 0, and if she wins her capital becomes 26^2, and she bets that $U_3 = x_3$. She continues till game 11, unless she has lost her capital earlier.

At time n the capital of this first gambler is

$$M_n = \begin{cases} 1 & \text{if } n = 0, \\ M_{n-1} 26 \mathbf{1}_{U_n = x_n} & \text{if } 1 \leq n \leq 11, \\ M_{11} & \text{if } n \geq 12. \end{cases}$$

The second gambler starts with capital 1, plays the same game with one time unit of delay, and bets on $U_2 = x_1$. If she loses, her capital at time 2 is 0, and she stops. If she wins, her capital at time 2 is 26, and she bets $U_3 = x_2$. At time 12 this gambler's capital is either 26^{11} or 0.

In general, for each $j \geq 1$, the jth gambler with an initial capital of 1 starts playing with $j - 1$ units of time delay, betting on the event $U_j = x_1$. As long as she keeps winning, she bets on $U_{j+1} = x_2, \ldots, U_{j+10} = x_{11}$. At time $j + 10$, the gambler stops playing.

Let $M_n^{(j)}$ be the capital of the jth gambler at time n. Then,

$$M_n^{(j)} = \begin{cases} 1 & \text{if } n < j, \\ M_{n-1}^{(j)} 26 \mathbf{1}_{U_n = x_{(n-j)+1}} & \text{if } j \leq n \leq j + 10, \\ M_{j+10}^{(j)} & \text{if } n > j + 10. \end{cases}$$

Define $N_0 = 0$, and the total net gain of all the gamblers equals

$$N_n = \sum_{j=1}^{n} (M_n^{(j)} - M_0^{(j)}) = \sum_{j=1}^{n} M_n^{(j)} - n. \tag{4.12}$$

We show that $(N_n, \mathcal{G}_n)$ is a martingale, and the OST can be applied. Further,

$$N_\tau = (26)^{11} + (26)^4 + 26 - \tau.$$

Note that $N_\tau = \sum_{j=1}^{\tau} M_\tau^{(j)} - \tau$, and $M_\tau^{(j)}$ is the capital at time τ of the jth gambler. We show that

$$M_\tau^{(j)} = \begin{cases} 26^{11} & \text{if } j = \tau - 10, \\ 26^4 & \text{if } j = \tau - 3, \\ 26 & \text{if } j = \tau \\ 0 & \text{if } j \notin \{\tau - 10, \tau - 3, \tau\}. \end{cases}$$

The complete word abracadabra first appears at time τ. Therefore, the $(\tau - 10)$th gambler has a capital of 26^{11}, that is, $M_\tau^{(\tau-10)} = 26^{11}$.

The $(\tau - 3)$th gambler has a capital of $M_\tau^{(\tau-3)} = 26^4$, because the last four letters of abracadabra coincide with its first four letters. Since the last letter is the same as the first letter, the τth gambler has won his first bet and his capital is 26. Finally, the jth gambler, $j \notin \{\tau - 10, \tau - 3, \tau\}$ have lost at least one bet, and hence their total capital is $M_\tau^{(j)} = 0$.

We now verify that for any fixed $j \in \mathbb{N}$, $(M_n^{(j)})_{n \geq 0}$ is a martingale. Note that $M_0^{(j)} = 1$ is $\mathcal{G}_0$-measurable. Further, $M_n^{(j)}$ is a measurable function of $M_{n-1}^{(j)}$ and U_n, assuming that $M_n^{(j-1)}$ is $\mathcal{G}_{n-1}$-measurable.

So it follows that $M_n^{(j)}$ is $\mathcal{G}_n$-measurable. Its integrability follows from the observation $|M_n^{(j)}| \leq 26|M_n^{(j-1)}| \leq \cdots (26)^n$.

It is easy to see that for $n < j$ or $n > j + 10$ we have

$$E(M_n^{(j)} | \mathcal{G}_{n-1}) = M_{n-1}^{(j)}.$$

For $n \in \{j, \ldots, j + 10\}$ we have

$$\begin{aligned} E(M_n^{(j)} | \mathcal{G}_{n-1}) &= E(M_{n-1}^{(j)} 26 \mathbf{1}_{U_n = x_{n-j+1}} | \mathcal{G}_{n-1}) \\ &= M_{n-1}^{(j)} 26 P(U_n = x_{n-j+1}) \\ &= M_{n-1}^{(j)}. \end{aligned}$$

Also note that for $n \in \{j, j+1, \ldots, j+10\}$ we have

$$\begin{aligned} |M_n^{(j)} - M_{n-1}^{(j)}| &\leq |26 \mathbf{1}_{U_n = x_{n-j+1}} - 1| |M_{n-1}^j| \\ &\leq 25|M_{n-1}^{(j)}|. \end{aligned}$$

Now since $M_{j-1}^{(j)} = 1$, it follows that

$$|M_n^{(j)} - M_{n-1}^{(j)}| \leq 25^{11}.$$

For $n < j$ and $n > j + 10$, the above difference is zero.

From above it is also clear that N_n, defined in (4.12) is $\mathcal{G}_n$-measurable and is integrable.

Now

$$E(N_n|\mathcal{G}_{n-1}) = \sum_{j=1}^{n} E(M_n^{(j)}|\mathcal{G}_{n-1})$$

$$= \sum_{j=1}^{n} M_{n-1}^{(j)} - n.$$

For $j = n$, $M_{n-1}^{(n)} = 1$. So the right side is $\sum_{j=1}^{n-1} M_{n-1}^{(j)} - (n-1)$. Hence we have

$$E(N_n|\mathcal{G}_{n-1}) = N_{n-1}.$$

This shows that $(N_n, \mathcal{G}_n)$ is a martingale.

Finally note that

$$N_n - N_{n-1} = \sum_{j=1}^{n} M_n^{(j)} - n - \sum_{j=1}^{n-1} M_{n-1}^{(j)} - (n-1)$$

$$\leq \sum_{j=n-10}^{n} |M_n^{(j)} - M_n^{(j-1)}|$$

$$\leq 11(25^{11}).$$

The result will then follow if we can show that $E(\tau) < \infty$. We leave this as an exercise. □

4.7 Exercises

Exercise 4.7.1 Let $(S_n)_{n \geq 0}$ be a non-symmetric random walk, that is, it takes a positive step and a negative step with respective probabilities p and $1 - p$, where $0 < p < 1$ and $p \neq \frac{1}{2}$. For integers $a < 0 < b$, calculate $E(\tau)$ where

$$\tau = \inf\{n \geq 1 : S_n \in \{a, b\}\}.$$

Exercise 4.7.2 Suppose $\{X_k\}$ are independent with $E(X_k) = 0$ and $|X_k| \leq K < \infty$ for all k. Let $S_n := \sum_{k=1}^{n} X_k$ and $s_n^2 := \sum_{k=1}^{n} E(X_k^2)$. Using a suitable stop time show that

$$P(\max_{1 \leq k \leq n} |S_k| \leq x) \leq (x + K)^2/s_n^2.$$

Exercise 4.7.3 The adapted sequence $(X_n, \mathcal{G}_n)_{n \geq 1}$ is called a **local martingale** if there exist a sequence of $\{\mathcal{G}_n\}$-stopping times $\{\tau_k\}$ such that $\tau_k \uparrow \infty$ almost surely and $(X_{n \wedge \tau_k}, \mathcal{G}_n)$ is a martingale for each k. Show that any martingale is a local martingale and any integrable, local martingale is a martingale.

Exercise 4.7.4 Verify the steps in the proof of Theorem 4.6.1 that have not been proved completely.

Exercise 4.7.5 Suppose $(X_n, \mathcal{G}_n)_{n\geq 1}$ is a martingale with $\mathrm{E}(X_n^2) < \infty$ for all n.
(a) Show that $(X_n^2, \mathcal{G}_n)_{n\geq 1}$ is a sub-martingale.
(b) Let

$$A_n = \sum_{i=2}^n \mathrm{E}((X_i - X_{i-1})^2 | \mathcal{G}_{i-1}), \ Y_n = X_n^2 - A_n, \ n \geq 1.$$

Verify that (Y_n, A_n) is the Doob decomposition of X_n^2 given by Theorem 4.1.1. The sequence $\{A_n\}$ is called the **quadratic variation** of $\{X_n\}$. This notion will be used later in Chaps. 7 and 8.

Exercise 4.7.6 Let $(X_k, \mathcal{G}_k)_{1\leq k\leq n}$ be a zero mean sub-martingale. Show that for $\lambda > 0$,

$$\lambda P\big(\max_{1\leq k\leq n} |S_k| > 2\lambda\big) \leq \lambda P(|S_n| > \lambda) + \mathrm{E}\big((|S_n| - 2\lambda)\mathbf{1}_{\{|S_n|\geq 2\lambda\}}\big)$$
$$\leq \mathrm{E}\big(|S_n|\mathbf{1}_{\{|S_n|>\lambda\}}\big).$$

For a solution see Page 16 of Hall and Heyde (1980).

References

F. Caravenna, The abracadabra problem (2016). https://staff.matapp.unimib.it/~fcaraven/download/other/abracadabra.pdf
J.L. Doob, *Classical Potential Theory and its Probabilistic Counterpart*, vol. 549 (Springer, 1984)
P. Hall, C.C. Heyde, *Martingale Limit Theory and its Application* (Academic, Harcourt Brace Jovanovich, Publishers, New York-London, 1980)
P. Lévy, *Théorie de l'Addition des Variables Aléatoires* (Gauthier-Villars, Paris, 1937)
J. Ville, Étude critique de la notion de collectif. *Monographies des Probabilitès*, no. 3 (Gauthier-Villars, Paris, 1939). http://eudml.org/doc/192893
A. Wald, On cumulative sums of random variables. Ann. Math. Stat. **15**(3), 283–296 (1944)
A. Wald, Some generalizations of the theory of cumulative sums of random variables. Ann. Math. Stat. **16**(3), 287–293 (1945)
D. Williams, *Probability with Martingales*. Cambridge Mathematical Textbooks (Cambridge University Press, Cambridge, 1991)

Chapter 5
Almost Sure and L^p Convergence

This chapter is devoted to answering questions on the almost sure and L^p convergence of martingales and sub-martingales. The crucial tools in this study are the Maximal inequality of Chap. 4, and the Upcrossing lemma that we shall prove shortly. We also introduce the extremely useful concept of time-reversed martingales, which are called reverse martingales, and prove some convergence results. These will be used later to prove many results, such as Kolmogorov's 0-1 law and the SLLN for U-statistics.

5.1 Upcrossing Lemma

Definition 5.1.1 (*Upcrossing*) Suppose $\{X_n\}_{n \geq 1}$ is a sequence of real-valued random variables, and $-\infty < \alpha < \beta < \infty$. Every time $\{X_n\}$ is larger than β after being below α, we say that an **upcrossing** of the interval $[\alpha, \beta]$ has occurred. Let

$$\eta_0 := 0,$$
$$\eta_k := \begin{cases} \inf\{n > \eta_{k-1} : X_n \geq \beta\} & \text{if } k \text{ is even,} \\ \inf\{n > \eta_{k-1} : X_n \leq \alpha\} & \text{if } k \text{ is odd.} \end{cases}$$

The **number of upcrossings of** $[\alpha, \beta]$ by $\{X_1, \ldots, X_n\}$ is defined as

$$U_n[\alpha, \beta] := \sup\{i \geq 0 : \eta_{2i} \leq n\}.$$

◇

See Fig. 5.1 for an illustration.

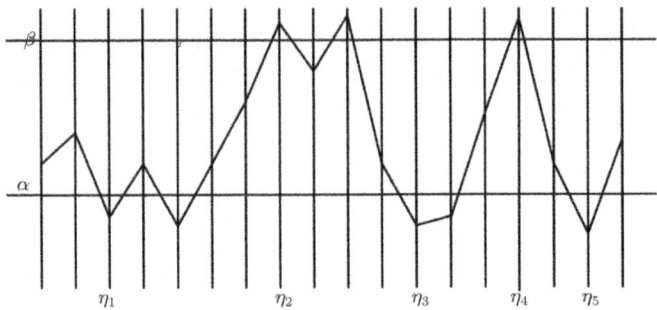

Fig. 5.1 Two upcrossings of $[\alpha, \beta]$

Lemma 5.1.1 (Upcrossing) *Let $(X_n, \mathcal{G}_n)_{1 \leq n < \infty}$ be a sub-martingale. Then*

$$\mathrm{E}(U_n[\alpha, \beta]) \leq \frac{\alpha + \mathrm{E}(|X_n|)}{\beta - \alpha} \text{ for every } n \geq 1.$$

♦

Proof Define

$$\theta := \beta - \alpha, \text{ and } Y_k := \max\{0, X_k - \alpha\}, \, k \geq 1.$$

Since $Y_k = \varphi(X_k)$ where φ is a convex function of X_k, and is integrable, by Exercise 4.1.3(b), $(Y_n, \mathcal{G}_n)$ is a sub-martingale. Define

$$\tau_0 := 0,$$
$$\tau_k := \begin{cases} \inf\{t > \tau_{k-1} : Y_t \geq \theta\} & \text{if } k \text{ is even,} \\ \inf\{t > \tau_{k-1} : Y_t = 0\} & \text{if } k \text{ is odd.} \end{cases}$$

Fix $n \geq 1$. Then $\{\tau_k \wedge n\}$ is a sequence of non-decreasing bounded stop times, and $n \wedge \tau_{k-1} < n \wedge \tau_k$ if $\tau_{k-1} < n$. Hence, $\tau_n \geq n$. Therefore,

$$Y_n = Y_{\tau_n \wedge n} \geq Y_{\tau_n \wedge n} - Y_{\tau_1 \wedge n} \text{ (since } \{Y_k\} \text{ are non-negative)}$$
$$= \sum_{k=2}^{n}(Y_{\tau_k \wedge n} - Y_{\tau_{k-1} \wedge n})$$
$$= \sum_{1} + \sum_{2},$$

where $\sum_i$, $i = 1, 2$ denote the sums over even and odd k, $2 \leq k \leq n$.

By the OST Theorem 4.3.2, for all k, $\mathrm{E}(Y_{\tau_k \wedge n}) \geq \mathrm{E}(Y_{\tau_{k-1} \wedge n})$. Hence

$$E(Y_n) \geq E\left(\sum_1\right)$$
$$= E\left(\sum_{k \leq n, \text{ even}} (Y_{\tau_k \wedge n} - Y_{\tau_{k-1} \wedge n})\right)$$
$$\geq E\left(\sum_{\substack{k \leq n, \text{ even} \\ \tau_k \leq n}} (Y_{\tau_k \wedge n} - Y_{\tau_{k-1} \wedge n})\right) \text{ (for even } k, Y_{\tau_k \wedge n} \geq Y_{\tau_{k-1} \wedge n})$$
$$= \theta E(U_n[\alpha, \beta]).$$

Thus,
$$(\beta - \alpha) E(U_n[\alpha, \beta]) \leq E(Y_n) \leq |\alpha| + E(|X_n|).$$

This completes the proof. □

5.2 L^1-Bounded and UI Sub-martingales

Theorem 5.2.1 (L^1-bounded sub-martingales converge almost surely) *Let $\{X_n\}$ be a sub-martingale such that $\sup_{n \geq 1} E(|X_n|) < \infty$. Then there is an integrable X such that $X_n \to X$ almost surely.* ♦

Proof Fix $\alpha < \beta$. Let $U_n[\alpha, \beta]$ be the number of upcrossings of $[\alpha, \beta]$ by $\{X_1, \ldots, X_n\}$, $1 \leq n < \infty$. Note that $\{U_n[\alpha, \beta]\}$ is a non-negative non-decreasing sequence. Let $U_\infty[\alpha, \beta]$ be its limit. By the Monotone convergence Theorem 1.5.1(ii) and the Upcrossing Lemma 5.1.1,

$$E(U_\infty[\alpha, \beta]) = \lim_{n \to \infty} E(U_n[\alpha, \beta])$$
$$\leq (\beta - \alpha)^{-1} \sup_n (|\alpha| + E|X_n|) < \infty.$$

Thus, for every $\alpha < \beta$, $U_\infty(\alpha, \beta) < \infty$ almost surely. It then follows that

$$P(U_\infty[\alpha, \beta] < \infty \text{ for all } \alpha, \beta \in \mathbb{Q}, \alpha < \beta) = 1.$$

This clearly implies that $\lim_{n \to \infty} X_n$ exists, almost surely.

Now define $X := \limsup_{n \to \infty} X_n$. Then $X_n \to X$ almost surely. Fatou's lemma (Theorem 1.5.1(i)) implies that

$$E(|X|) \leq \liminf_{n \to \infty} E(|X_n|) < \infty,$$

and hence X is integrable. This completes the proof. □

Exercise 5.2.1 Theorem 5.2.1 fails without the L^1 boundedness. For the simple symmetric random walk martingale S_n, show that

$\sup_n E(|S_n|) = \infty$, and $\liminf_{n \to \infty} S_n = -\infty$, $\limsup_{n \to \infty} S_n = \infty$ almost surely.

Remark 5.2.1 The convergence in Theorem 5.2.1 need not hold in L^1. In the doubling game Example 4.1.2, let X_0 be the initial fixed asset. Then the non-negative martingale $\{X_n\}$ satisfies $E(X_n) = X_0$ for all n. Hence Theorem 5.2.1 applies, and indeed it is easy to see that $X_n \to 0$ almost surely, but obviously not in L^1. ●

Convergence in L^1 can be achieved if we add the UI condition.

Theorem 5.2.2 (Convergence of UI sub-martingales) *If $(X_n, \mathcal{G}_n)$ is a UI submartingale, then there exists a random variable X_∞ such that as $n \to \infty$, $X_n \to X_\infty$, almost surely, and in L^1. Furthermore,*

$$X_n \leq E(X_\infty | \mathcal{G}_n) \text{ almost surely for all } n \geq 1, \tag{5.1}$$

so that with $\mathcal{G}_\infty = \bigvee_{n \geq 1} \mathcal{G}_n$, $(X_n, \mathcal{G}_n)_{0 \leq n \leq \infty}$ is a sub-martingale. ◆

Proof Since $\{X_n\}$ is UI, it follows that $\sup_{n \geq 1} E(|X_n|) < \infty$. Using Theorem 5.2.1, there exists an integrable X_∞, such that $X_n \to X_\infty$ almost surely. Hence, by uniform integrability, $X_n \to X_\infty$ in L^1. Fix $n \in \mathbb{N}$. It is then immediate that

$$E(X_m | \mathcal{G}_n) \to E(X_\infty | \mathcal{G}_n) \text{ in } L^1 \text{ as } m \to \infty.$$

Since for all $m \geq n$,
$$X_n \leq E(X_m | \mathcal{G}_n) \text{ almost surely,}$$

(5.1) follows, and this completes the proof. □

If $\{X_n\}$ is a martingale, then $\{-X_n\}$ is also so. Hence the following theorem is an immediate corollary of Theorem 5.2.2.

Theorem 5.2.3 (L^1 convergence of UI martingales) *Let $(X_n, \mathcal{G}_n)$ be a martingale which is UI. Then there exists an integrable X_∞ such that, almost surely and in L^1, $X_n \to X_\infty$. Furthermore,*

$$X_n = E(X_\infty | \mathcal{G}_n) \text{ almost surely for all } n \geq 1, \tag{5.2}$$

so that with $\mathcal{G}_\infty = \bigvee_{n \geq 1} \mathcal{G}_n$, $(X_n, \mathcal{G}_n)_{0 \leq n \leq \infty}$ is a martingale. ◆

Note that relation (5.2) says that $(X_n, \mathcal{G}_n)$ is a Doob's martingale. The following result is a consequence of the above.

Theorem 5.2.4 (Lévy's upward theorem) *Let $\{\mathcal{G}_n\}$ be a filtration, and X be an integrable random variable. Then as $n \to \infty$,*

$$E(X | \mathcal{G}_n) \to E(X | \bigvee_{n=1}^{\infty} \mathcal{G}_n), \text{ almost surely, and in } L^1.$$

◆

Proof Let

$$X_n := \mathrm{E}(X|\mathcal{G}_n), \ n \geq 1, \ \mathcal{G}_\infty = \bigvee_{n=1}^\infty \mathcal{G}_n.$$

Then $(X_n, \mathcal{G}_n)$ is a Doob's martingale.

By Theorem 3.4.3, $\{X_n\}$ is UI. Hence, by Theorem 5.2.3, there exists X_∞ such that as $n \to \infty$,

$$X_n \to X_\infty \text{ almost surely, and in } L^1.$$

Since each X_n is $\mathcal{G}_\infty$-measurable, redefining

$$X_\infty = \limsup_{n \to \infty} X_n,$$

it follows that X_∞ is $\mathcal{G}_\infty$-measurable.

Fix $A \in \cup_{n=1}^\infty \mathcal{G}_n$. Then, for some n_0, $A \in \mathcal{G}_{n_0} \subset \mathcal{G}_n$ for all $n \geq n_0$. So,

$$\int_A X_n \, dP = \int_A X \, dP. \tag{5.3}$$

Since $X_n \to X_\infty$ in L^1, by taking limit in (5.3) it is also true that

$$\int_A X_\infty \, dP = \int_A X \, dP \text{ for all } A \in \cup_{n=1}^\infty \mathcal{G}_n \tag{5.4}$$

The class of sets A for which (5.4) holds, includes the field $\cup_{n=1}^\infty \mathcal{G}_n$, and at the same time is also a monotone class. So, by the Monotone class Theorem 1.1.1, it includes $\sigma(\cup_{n=1}^\infty \mathcal{G}_n) = \mathcal{G}_\infty$. That is, for all $A \in \mathcal{G}_\infty$,

$$\int_A X_\infty \, dP = \int_A X \, dP \text{ for all } A \in \mathcal{G}_\infty.$$

In other words, $X_\infty = \mathrm{E}(X|\mathcal{G}_\infty)$. Hence the proof is complete. □

5.3 L^p Convergence for $p > 1$

Theorem 5.3.1 (Convergence of an L^p-bounded martingale) *Let* $(X_n, \mathcal{G}_n)_{1 \leq n < \infty}$ *be a martingale such that for some* $1 < p < \infty$,

$$\sup_{n \geq 1} \mathrm{E}(|X_n|^p) < \infty. \tag{5.5}$$

Then there exists $X_\infty \in L^p$ such that, as $n \to \infty$, $X_n \to X_\infty$ almost surely and in L^p. Moreover,

$$X_n = E(X_\infty | \mathcal{G}_n) \text{ almost surely for all } n \geq 1. \tag{5.6}$$

♦

Proof Condition (5.5) implies that the martingale $\{X_n\}$ is UI (see Exercise 3.5.23). Hence by Theorem 5.2.3, there exists X_∞ such that, as $n \to \infty$, $X_n \to X_\infty$ almost surely and in L^1. In addition, (5.6) holds.

To prove convergence in L^p using Theorem 3.4.2, it is sufficient to show that $\{|X_n|^p\}$ is UI. We show this using Corollary 4.5.1 of Doob's Maximal Inequality Theorem 4.5.2.

Note that $(|X_n|^p, \mathcal{G}_n)_{n \geq 1}$ is a sub-martingale. Now,

$$E\left(\max_{1 \leq i \leq n} |X_i|^p\right)$$

$$= \int_0^\infty P\left(\max_{1 \leq i \leq n} |X_i|^p \geq x\right) dx \quad \text{(by Exercise 1.10.2)}$$

$$= \int_0^\infty P\left(\max_{1 \leq i \leq n} |X_i| \geq x^{1/p}\right) dx$$

$$= p \int_0^\infty z^{p-1} P\left(\max_{1 \leq i \leq n} |X_i| \geq z\right) dz \quad \text{(change of variable)}$$

$$\leq p \int_0^\infty z^{p-2} \int_{\{\max_{1 \leq i \leq n} |X_i| \geq z\}} |X_n| dP \, dz \quad \text{(Theorem 4.5.2)}$$

$$= p \int_\Omega |X_n| \int_0^{\max_{1 \leq i \leq n} |X_i|} z^{p-2} \, dz \, dP \quad \text{(Fubini's theorem)}$$

$$= \frac{p}{p-1} \int_\Omega |X_n| \left(\max_{1 \leq i \leq n} |X_i|\right)^{p-1} dP$$

$$= q E\left(|X_n| \max_{1 \leq i \leq n} |X_i|^{p-1}\right) \quad \text{(where } q = \frac{p}{p-1}\text{)}$$

$$\leq q \left(E(|X_n|^p)\right)^{1/p} \left(E\left(\max_{1 \leq i \leq n} |X_i|^{(p-1)q}\right)\right)^{1/q},$$

the last line following by Hölder's inequality.

Upon adjustment, the above inequality is the same as

$$E\left(\max_{1 \leq i \leq n} |X_i|^p\right) \leq q^p E(|X_n|^p). \tag{5.7}$$

Letting $n \to \infty$,

$$E\left(\sup_{n \geq 1} |X_n|^p\right) \leq q^p \sup_{n \geq 1} E(|X_n|^p) < \infty.$$

This implies that $\{|X_n|^p\}$ is UI, and hence the proof is complete. □

5.4 Reverse Martingale

Incidentally, (5.7) is also known as *Doob's maximal inequality*. Note that this inequality cannot be true for $p = 1$ in general. As we have already seen, Theorem 5.3.1 is also *not true* for $p = 1$.

Example 5.3.1 (*Pólya's urn*) Consider an urn which has black and white balls, where the initial proportion of white balls is X_0. A ball is drawn at random and replaced. An additional ball of the same color is added to the urn. If X_n denotes the proportion of white balls after n draws, then we have seen in Example 4.1.6 that $\{X_n\}$ is a martingale, bounded below and above by 0 and 1.

By Theorem 5.3.1, there exists a random variable X_∞ ($0 \leq X_\infty \leq 1$), such that for all $1 \leq p < \infty$, as $n \to \infty$, $X_n \to X_\infty$ almost surely and in L^p, and $\mathrm{E}(X_\infty) = X_0$. On the other hand $\{X_n^2\}$ is a sub-martingale. Hence it follows that

$$\mathrm{Var}(X_n) = \mathrm{E}(X_n^2) - X_0^2 \uparrow \mathrm{E}(X_\infty^2) - X_0^2 = \mathrm{Var}(X_\infty).$$

If $0 < X_0 < 1$, then $\mathrm{Var}(X_1) > 0$ and hence $\mathrm{Var}(X_\infty) > 0$. That is, the limit random variable X_∞ is *not* a constant. ▲

5.4 Reverse Martingale

The idea of a reverse martingale is based on time reversal. Analogous to the results on convergence of martingales, we shall establish convergence of reverse martingales. This in turn will be used later to prove many results, such as Kolmogorov's 0-1 law and the SLLN for U-statistics.

Definition 5.4.1 (*Reverse filtration and martingale*) Let $(\Omega, \mathcal{A}, P)$ be a probability space.

(a) A family $\{\mathcal{G}_n\}_{n \geq 1}$ of σ-fields is said to be a **reverse filtration** if $\mathcal{A} \supset \mathcal{G}_1 \supset \mathcal{G}_2 \supset \cdots$.

(b) Suppose $\{\mathcal{G}_n\}_{n \geq 1}$ is a reverse filtration, and for each n, X_n is integrable, and $\mathcal{G}_n$-measurable. Then $(X_n, \mathcal{G}_n)_{n \geq 1}$ is said to be a **reverse martingale** if for all $n \in \mathbb{N}$

$$\mathrm{E}(X_{n-1} | \mathcal{G}_n) = X_n, \ n \geq 2.$$

Reverse sub- and super-martingales are defined in a similar manner. ◇

Exercise 5.4.1 Suppose $(X_n, \mathcal{G}_n)_{n \geq 1}$ is a reverse martingale. Then show that for every n, the sequence $(X_n, \mathcal{G}_n), (X_{n-1}, \mathcal{G}_{n-1}), \ldots, (X_1, \mathcal{G}_1)$ is a martingale, and $\{\mathrm{E}|X_n|\}_{n \geq 1}$ is non-increasing.

Theorem 5.4.1 (Convergence of reverse martingale) *If* $(X_n, \mathcal{G}_n), n \geq 1$ *is a reverse martingale, then as* $n \to \infty$,

$$X_n \to \mathrm{E}\Big(X_1 \Big| \bigcap_{n=1}^{\infty} \mathcal{G}_n\Big), \quad \text{almost surely and in } L^1.$$

♦

Proof For $-\infty < \alpha < \beta < \infty$, let $U_n[\alpha, \beta]$ be the number of upcrossings of $[\alpha, \beta]$, by $\{X_n, \ldots, X_1\}$. Since $(X_n, \mathcal{G}_n), (X_{n-1}, \mathcal{G}_{n-1}), \ldots, (X_1, \mathcal{G}_1)$ is a martingale, by Lemma 5.1.1,

$$\mathrm{E}(U_n[\alpha, \beta]) \leq \frac{|\alpha| + \mathrm{E}(|X_1|)}{\beta - \alpha}.$$

Since the right side is independent of n, as in the proof of Theorem 5.2.1, it follows that X_n converges almost surely to a random variable X_∞. We can re-define X_∞ as

$$X_\infty := \limsup_{n \to \infty} X_n.$$

Then clearly X_∞ is measurable with respect to $\cap_{n=1}^{\infty} \mathcal{G}_n$. Note that

$$\mathrm{E}(|X_\infty|) \leq \limsup \mathrm{E}(|X_n|) \leq \mathrm{E}(|X_1|). \tag{5.8}$$

Since $X_n = \mathrm{E}(X_1|\mathcal{G}_n)$, it is UI. Hence, for any $A \in \cap_{n=1}^{\infty} \mathcal{G}_n$,

$$\int_A X_\infty dP = \lim_{n \to \infty} \int_A X_n dP$$
$$= \int_A X_1 dP \quad \text{(by reverse martingale property)}.$$

Thus, $\mathrm{E}\big(X_1|\cap_{n=1}^{\infty} \mathcal{G}_n\big) = X_\infty$, and the proof is complete. □

Lévy's downward theorem is really a restatement of the above result.

Theorem 5.4.2 (Lévy's downward theorem) *Suppose Z is any integrable random variable and $\{\mathcal{G}_n\}_{n \geq 1}$ is a reverse filtration. Then,*

$$\lim_{n \to \infty} \mathrm{E}(Z|\mathcal{G}_n) = \mathrm{E}\Big(Z \Big| \bigcap_{n=1}^{\infty} \mathcal{G}_n\Big) \quad \text{almost surely and in } L^1.$$

♦

We can combine the two Lévy's Theorems 5.2.4 and 5.4.2 into one convenient statement.

Theorem 5.4.3 (Lévy's theorem) *Let Z be an integrable random variable and $\{\mathcal{G}_n\}$ be either a filtration or a reverse filtration. Then,*

$$\lim_{n \to \infty} \mathrm{E}(Z|\mathcal{G}_n) = \mathrm{E}(Z|\mathcal{G}_\infty) \quad \text{almost surely and in } L^1,$$

where

$$\mathcal{G}_\infty := \begin{cases} \bigvee_{n=1}^\infty \mathcal{G}_n & \text{if } \{\mathcal{G}_n\} \text{ is a filtration,} \\ \bigcap_{n=1}^\infty \mathcal{G}_n & \text{if } \{\mathcal{G}_n\} \text{ is a reverse filtration.} \end{cases}$$

♦

5.5 Exercises

Exercise 5.5.1 Prove the following inequality, which is more general than the Upcrossing Lemma 5.1.1. Suppose $(X_k, \mathcal{G}_k)_{1 \leq k \leq n}$ is a sub-martingale. Let $U_n[\alpha, \beta]$ be the number of upcrossings of the interval $[\alpha, \beta]$ by $\{X_k\}_{1 \leq k \leq n}$. Then

$$\mathrm{E}(U_n[\alpha, \beta]) \leq \frac{\mathrm{E}((X_n - \alpha)^+) - \mathrm{E}((X_1 - a)^+)}{\beta - \alpha}.$$

Exercise 5.5.2 Prove or give counter-example to the following.

(a) Let $\{X_n\}_{n \geq 1}$ be a martingale and $X_n \to X_\infty$ in L^1 as $n \to \infty$. If X_n is not a constant for some $n \in \mathbb{N}$, then X_∞ is also not a constant.

(b) Let $\{X_n\}_{n \geq 1}$ be a martingale such that as $n \to \infty$, $X_n \to X_\infty$ a.s. If X_n is not a constant for some $n \in \mathbb{N}$, then X_∞ is also not a constant.

Exercise 5.5.3 Let $(\Omega, \mathcal{A}, \{\mathcal{G}_n\}_{n \geq 1}, P)$ be a filtered probability space with $\mathcal{A} = \bigvee_{n=1}^\infty \mathcal{G}_n$. Let μ be a finite measure on $(\Omega, \mathcal{A})$ such that for $n \geq 1$, on $\mathcal{G}_n$, $\mu \ll P$, with $\dfrac{d\mu}{dP} = X_n$.

(a) Show that there exists an integrable X_∞ such that as $n \to \infty$, $X_n \to X_\infty$ almost surely.

(b) Prove that the following are equivalent:
   (i) $\mathrm{E}(X_\infty) = \mu(\Omega)$,
   (ii) $\mu \ll P$ on $\mathcal{A}$,
   (iii) $\{X_n\}_{1 \leq n < \infty}$ is UI.

Exercise 5.5.4 Let $\{X_n\}_{n \geq 1}$ be a non-negative super-martingale.

(a) Show that there exists an integrable random variable $X_\infty \geq 0$ such that as $n \to \infty$, $X_n \to X_\infty$ almost surely.

(b) Show that $\mathrm{E}(X_\infty) \geq \mathrm{E}(X_1)$ if and only if $\{X_n\}_{n \geq 1}$ is a UI martingale.

Exercise 5.5.5 Suppose $\{Y_i\}$ are independent and non-negative, with $\mathrm{E}(Y_n) = 1$ for all n. Let $X_n := Y_1 \cdots Y_n$. Show that X_n converges almost surely to an integrable random variable.

Exercise 5.5.6 Suppose $(X_n, \mathcal{G}_n)_{n\geq 1}$ is a martingale such that

$$\sum_{n=1}^{\infty} E\big((X_{n+1} - X_n)^2\big) < \infty. \tag{5.9}$$

(a) Show that X_n converges almost surely.

(b) Show that (5.9) is equivalent to $E(A_\infty) < \infty$ where

$$A_n = \sum_{i=2}^{n} E((X_i - X_{i-1})^2 | \mathcal{G}_{i-1}), \, n \geq 1,$$

is the quadratic variation of (X_n) and $A_\infty = \lim_{n\to\infty} A_n$.

(c) If $(X_{n+1} - X_n)^2$ in (5.9) is replaced by $|X_{n+1} - X_n|^p$ for some $p > 2$, then does (a) remain true?

Exercise 5.5.7 Let $(X_n, \mathcal{G}_n)$ be a martingale which is bounded, either above or below. Show that $\sup_{n\geq 1} E|X_n| < \infty$.

Exercise 5.5.8 Suppose $(X_n, \mathcal{G}_n)$ is an L^1-bounded martingale. Show that there exists non-negative martingales $(Y_n, \mathcal{G}_n)$ and $(Z_n, \mathcal{G}_n)$ such that $X_n = Y_n - Z_n$ for all $n \geq 1$.

Exercise 5.5.9 (*This requires familiarity with the notions of recurrence and irreducibility of a countable Markov chain. See* Norris (1998) *for these notions.*) Let $\{X_n\}_{n\geq 1}$ be a Markov chain with state space S and transition probability matrix P. Suppose that $\phi : S \to \mathbb{R}$ is a bounded function such that for all $i \in S$,
$$\phi(i) \geq \sum_{j \in S} P(i, j)\phi(j).$$

(a) Show that $\phi(X_n)$ converges almost surely.

(b) If the chain is irreducible and recurrent, then show that ϕ has to be a constant function.

Exercise 5.5.10 Give an alternative proof of Theorem 5.3.1 by proceeding as follows. Let $(X_n, \mathcal{G}_n)$ satisfy the hypotheses of Theorem 5.3.1.

(a) Show that $\{X_n : n \geq 1\}$ is UI.

(b) Prove that there exists $X_\infty \in L^1$ such that $X_n \to X_\infty$ almost surely, and we also have $X_n = E(X_\infty | \mathcal{G}_n)$ for all $n \geq 1$.

(c) Use Fatou's lemma to prove $X_\infty \in L^p$.

(d) Show that $|X_n|^p \leq E(|X_\infty|^p | \mathcal{G}_n), \, n \geq 1$.

(e) Prove that $\{|X_n|^p : n \geq 1\}$ is UI and thereby complete the proof of Theorem 5.3.1.

5.5 Exercises

Exercise 5.5.11 (*Theorem 2 of* Blackwell and Dubins (1971)) Let $\{X_n\}$ be random variables on $(\Omega, \mathcal{A}, P)$ such that $E(\sup_{n\geq 1} |X_n|) < \infty$ and $X_n \to X$ almost surely. Let $\{\mathcal{G}_n\}$ be sub-σ-fields which are either non-increasing or non-decreasing.

Show that in either case there exists σ-field $\mathcal{G}$ (identify $\mathcal{G}$ in the two cases) such that
$$\lim_{\substack{j \to \infty \\ n \to \infty}} E(X_n | \mathcal{G}_j) = E(X|\mathcal{G}) \text{ almost surely.}$$

Exercise 5.5.12 Let $\{X_n\}$ be a martingale such that
$$\sup_n |X_{n+1} - X_n| \leq K, \text{ for some constant } 0 < K < \infty. \tag{5.10}$$

Show that $P(A \cup B) = 1$, where
$$A = \{\omega : \lim_{n \to \infty} X_n(\omega) \text{ exists and is finite}\}$$
$$B = \{\omega : \limsup_{n \to \infty} X_n(\omega) = \infty \text{ and } \liminf_{n \to \infty} X_n(\omega) = -\infty\}.$$

Exercise 5.5.13 Suppose $(X_n, \mathcal{G}_n)$ is an adapted sequence such that, $X_0 = 0$ and $\{X_n\}$ satisfies (5.10). Suppose in addition, there are positive constants a and b such that
$$E(X_{n+1} - X_n + a \mid \mathcal{G}_n) I_{\{X_n \geq b\}} \leq 0, \text{ for all } n \geq 0.$$
Let $N_n = \sum_{k=1}^n \mathbf{1}_{\{X_k < b\}}$. Then show that $P(N_\infty < \infty) = 0$.

Exercise 5.5.14 Let $\{\mathcal{G}_n\}$ be a filtration and $\{X_n\}$, $\{\beta_n\}$ and $\{Y_n\}$ be positive random variables adapted to $\mathcal{G}_n$. Suppose
$$E(X_{n+1} \mid \mathcal{G}_n) \leq (1 + \beta_n)X_n + Y_n \text{ for all } n \geq 1.$$

Then $(X_n, \mathcal{G}_n)$ is called an **almost super-martingale**. Robbins and Siegmund (1971) proved and applied this result to study properties of certain stochastic approximation algorithms. Show that if $(X_n, \mathcal{G}_n)$ is an almost super-martingale, then X_n converges almost surely on the set
$$A = \{\sum_{n=1}^\infty \beta_n < \infty, \sum_{n=1}^\infty Y_n < \infty\}.$$

Hint: Define
$$X'_n = \frac{X_n}{(1+\beta_1)(1+\beta_2)\cdots(1+\beta_{n-1})}$$
$$Y'_n = \frac{Y_n}{(1+\beta_1)(1+\beta_2)\cdots(1+\beta_{n-1})}$$

$$U_n = X_n - \sum_{m=1}^{n-1} Y'_m$$

$$\tau_a = \inf\{n : \sum_{m=1}^{n} Y'_m > a\}.$$

Check and use the fact that $a + U_{\tau_a \wedge n}$ is a positive super-martingale.

References

D. Blackwell, L. Dubins, Merging of opinions with increasing information. Ann. Math. Stat. **33**(3), 882–886 (1962)

J.R. Norris, *Markov Chains*, in volume 2 of *Cambridge Series in Statistical and Probabilistic Mathematics* (Cambridge University Press, Cambridge, 1998). Reprint of 1997 original

H. Robbins, D. Siegmund, A convergence theorem for nonnegative almost supermartingales and some applications, in *Optimizing Methods in Statistics*, ed. by J.S. Rustagi (Academic Press, 1971), pp. 233–257

Chapter 6
Application of Convergence Theorems

We will now show several applications of the convergence results for martingales and reverse martingales. These include Kolmogorov's 0-1 law, strong law for iid sequences, strong law for U-statistics, Hewitt–Savage 0-1 law, strong law and de Finetti's theorem for exchangeable sequences, and Kakutani's theorem for likelihood ratios.

6.1 Kolmogorov's 0-1 Law

Definition 6.1.1 (*Tail σ-field*) Let $(\Omega, \mathcal{A})$ be a measurable space. For sub-σ-fields $\{\mathcal{G}_n\}$, its **tail σ-field** is defined as

$$\mathcal{T} := \bigcap_{n=1}^{\infty} \bigvee_{i=n}^{\infty} \mathcal{G}_i.$$

◇

Exercise 6.1.1 Suppose $(\Omega, \mathcal{A}, P)$ is a probability space. Show that any σ-field $\mathcal{G} \subset \mathcal{A}$ is trivial if and only if any $\mathcal{G}$-measurable random variable is a constant almost surely.

Theorem 6.1.1 (Kolmogorov's 0-1 law) Suppose $(\Omega, \mathcal{A}, P)$ is a probability space. Suppose $\{\mathcal{G}_n\}$ is a sequence of independent sub-σ-fields of $\mathcal{A}$, and $\mathcal{T}$ is its tail σ-field. Then, $\mathcal{T}$ is a trivial σ-field. ◆

Proof For $n \in \mathbb{N}$, define the σ-fields,

$$\mathcal{K}_n := \bigvee_{i=1}^{n} \mathcal{G}_i, \quad \mathcal{T}_n := \bigvee_{i=n+1}^{\infty} \mathcal{G}_i.$$

Fix $A \in \mathcal{T}$. Then $A \in \mathcal{T}_n$ for every $n \in \mathbb{N}$. Since for every n, $\mathcal{K}_n$ and $\mathcal{T}_n$ are independent, we have
$$\mathrm{E}\,(\mathbf{1}_A|\mathcal{K}_n) = \mathrm{E}(\mathbf{1}_A) = P(A). \tag{6.1}$$

Since $\{\mathcal{K}_n\}$ is a filtration and $\bigvee_{n=1}^{\infty} \mathcal{K}_n = \bigvee_{i=1}^{\infty} \mathcal{G}_i$, by Lévy's upward Theorem 5.2.4, as $n \to \infty$,
$$\mathrm{E}\,(\mathbf{1}_A|\mathcal{K}_n) \to \mathrm{E}\big(\mathbf{1}_A\big| \bigvee_{i=1}^{\infty} \mathcal{G}_i\big) \text{ almost surely.} \tag{6.2}$$

But then $\bigvee_{i=1}^{\infty} \mathcal{G}_i \supset \mathcal{T}$, and hence it follows that
$$\mathrm{E}\big(\mathbf{1}_A\big| \bigvee_{i=1}^{\infty} \mathcal{G}_i\big) = \mathbf{1}_A, \text{ almost surely.} \tag{6.3}$$

Using (6.1), (6.2) and (6.3), $P(A) = \mathbf{1}_A$ almost surely. Hence $P(A)$ is 0 or 1, and the proof is complete. $\square$

6.2 Strong Law for iid Sequences

Theorem 6.1.1 along with the reverse martingale convergence Theorem 5.4.1 implies the SLLN in the iid case. We need the following result.

Lemma 6.2.1 *Let X be an integrable random variable on $(\Omega, \mathcal{A}, P)$. Suppose $\mathcal{D}$ and $\mathcal{G}$ are sub-σ-fields of $\mathcal{A}$ such that $\mathcal{G} \vee \sigma(X)$ is independent of $\mathcal{D}$. Then*
$$\mathrm{E}\,(X|\mathcal{D} \vee \mathcal{G}) = \mathrm{E}\,(X|\mathcal{G}) \text{ almost surely.}$$

♦

Proof Let $Y := \mathrm{E}\,(X|\mathcal{G})$. Clearly, Y is $\mathcal{D} \vee \mathcal{G}$-measurable. It suffices to show that
$$\mathrm{E}\,(Y\mathbf{1}_E) = \mathrm{E}(X\mathbf{1}_E) \text{ for all } E \in \mathcal{D} \vee \mathcal{G}. \tag{6.4}$$

Let
$$\mathcal{S} := \{A \cap B : A \in \mathcal{D}, \ B \in \mathcal{G}\}.$$

We now show that $\mathcal{S}$ is a semi-field.

Suppose $E, D \in \mathcal{S}$. Then clearly $E \cap D \in \mathcal{S}$. Write $E = A \cap B$ for $A \in \mathcal{D}$, $B \in \mathcal{G}$. Then
$$E^c = A^c \cup (A \cap B^c),$$
and A^c, $A \cap B^c$ are disjoint and belong to $\mathcal{S}$. Thus, $\mathcal{S}$ is a semi-field.

6.2 Strong Law for iid Sequences

Moreover, for E as above, $X\mathbf{1}_B$ and $\mathbf{1}_A$ are $\sigma(X) \vee \mathcal{G}$ and $\mathcal{D}$-measurable respectively. Hence they are independent.
Therefore,

$$\begin{aligned}
\mathrm{E}(X\mathbf{1}_E) &= \mathrm{E}\big((X\mathbf{1}_B)\mathbf{1}_A\big) \\
&= \mathrm{E}(X\mathbf{1}_B)\mathrm{E}(\mathbf{1}_A) \text{ (by independence)} \\
&= \mathrm{E}(Y\mathbf{1}_B)\mathrm{E}(\mathbf{1}_A) \text{ (since } B \in \mathcal{G} \text{ and } Y = \mathrm{E}(X|\mathcal{G})) \\
&= \mathrm{E}(Y\mathbf{1}_B\mathbf{1}_A) \text{ (since } \mathcal{G} \text{ and } \mathcal{D} \text{ are independent)} \\
&= \mathrm{E}(Y\mathbf{1}_E).
\end{aligned}$$

Thus, (6.4) holds for all $E \in \mathcal{S}$.

Since $\mathcal{S}$ is a semi-field, by Exercise 1.2.1(b), the smallest field $\mathcal{F}(\mathcal{S})$ containing $\mathcal{S}$ consists of disjoint unions of sets from $\mathcal{S}$. Hence by additivity of expectations, (6.4) holds for all sets $E \in \mathcal{F}(\mathcal{S})$.

Now observe that the class of sets E for which (6.4) holds, is a monotone class. The Monotone class Theorem 1.1.1 implies that (6.4) holds for all $E \in \sigma(\mathcal{S})$. Now, note that $\sigma(\mathcal{S}) = \mathcal{D} \vee \mathcal{G}$. This completes the proof. □

Exercise 6.2.1 Let $\{X_i\}$ be iid random variables and $\mathrm{E}(|X_1|) < \infty$. Let $S_n = \sum_{i=1}^n X_i$. Show that $\mathrm{E}(X_k|S_n) = \frac{S_n}{n}$ for all $k \leq n$.

Theorem 6.2.1 (Strong law of large numbers (SLLN)) *Suppose $\{X_n\}_{n\geq 1}$ are iid with finite mean μ. Let $S_n := \sum_{i=1}^n X_i$, $n \geq 1$. Then,*

$$\lim_{n \to \infty} n^{-1} S_n = \mu \text{ almost surely, and in } L^1.$$

♦

Proof Define the reverse filtration $\{\mathcal{D}_n\}$ as

$$\begin{aligned}
\mathcal{D}_n &:= \sigma(S_n, S_{n+1}, S_{n+2}, \ldots) \\
&= \sigma(S_n, X_{n+1}, X_{n+2}, \ldots) \\
&= \sigma(S_n) \vee \sigma(X_{n+1}, X_{n+2}, \ldots).
\end{aligned}$$

Fix $n \geq 1$. Let
$$\mathcal{D} := \sigma(X_{n+1}, X_{n+2}, \ldots) \text{ and } \mathcal{G} := \sigma(S_n).$$

Clearly, $\sigma(X_1) \vee \mathcal{G}$ and $\mathcal{D}$ are independent. Hence by Lemma 6.2.1,

$$\mathrm{E}(X_1|\mathcal{D} \vee \mathcal{G}) = \mathrm{E}(X_1|\mathcal{G}),$$

that is,

$$\mathrm{E}(X_1|\mathcal{D}_n) = \mathrm{E}(X_1|S_n) = \frac{S_n}{n} \text{ (by Exercise (6.2.1))}.$$

Since $\{\mathcal{D}_n\}$ is a reverse filtration, by Theorem 5.4.1, as $n \to \infty$,

$$n^{-1} S_n = \mathrm{E}(X_1 | \mathcal{D}_n) \to \mathrm{E}\bigl(X_1 \bigm| \cap_{n=1}^{\infty} \mathcal{D}_n\bigr) \text{ almost surely and in } L^1.$$

Let $Z := \limsup_{n \to \infty} n^{-1} S_n$.

It is immediate that Z is $\sigma(X_k, X_{k+1}, \ldots)$-measurable for all $k \geq 1$. Therefore, Z is measurable with respect to

$$\mathcal{T} = \bigcap_{k=1}^{\infty} \sigma(X_k, X_{k+1}, \ldots).$$

Kolmogorov's 0-1 law Theorem 6.1.1 implies that $\mathcal{T}$ is a trivial σ-field, and hence Z is a constant almost surely. But since

$$Z = \mathrm{E}\bigl(X_1 \bigm| \bigcap_{n=1}^{\infty} \mathcal{D}_n\bigr) \text{ almost surely},$$

it follows that $Z = \mathrm{E}(X_1) = \mu$ almost surely. Thus, we have shown that

$$\lim_{n \to \infty} n^{-1} S_n = \mu \text{ almost surely, and in } L^1.$$

$\square$

6.3 Strong Law for U-Statistics

U-statistics arises from certain optimality considerations in statistics. It extends the idea of taking the sample average, by taking averages of symmetric functions of the observations. Their behavior is crucial in deriving properties of statistical estimates. Let $\{X_i\}$ be iid random variables taking values in some topological space S. Let $h(x_1, \ldots, x_m)$ be any real-valued Borel measurable function on S^m which is *symmetric in its arguments*—for every permutation π of $\{1, \ldots, m\}$, $h(x_1, \ldots, x_m) = h(x_{\pi(1)}, \ldots, x_{\pi(m)})$. Such a function h will be referred to as a **kernel**.

Definition 6.3.1 (U-*statistics*) The U-statistic of *order* or *degree* m, with *kernel* h is

$$U_n(h) := \binom{n}{m}^{-1} \sum_{1 \leq i_1 < \cdots < i_m \leq n} h(X_{i_1}, \ldots, X_{i_m}) \tag{6.5}$$

$$= \frac{1}{m!} \binom{n}{m}^{-1} \sum_{\substack{1 \leq i_1, \ldots, i_m \leq n, \\ i_j \neq i_k \text{ for all } j \neq k}} h(X_{i_1}, \ldots, X_{i_m}).$$

6.3 Strong Law for U-Statistics

We write U_n for $U_n(h)$. Note that the summands in (6.5) have identical probability distribution. The simplest examples of U-statistics are:

$$h(x) = x, \quad (m = 1, \text{ the sample mean})$$
$$h(x_1, x_2) = (x_1 - x_2)^2/2 \ (m = 2, \text{ the sample variance}).$$

Clearly, if $E(|h(X_1, \ldots, X_m)|) < \infty$, then $E(U_n) = E(h(X_1, \ldots, X_m))$.

The SLLN for U-statistics extends Theorem 6.2.1. Its proof by Berk (1966) uses the reverse martingale convergence Theorem 5.4.1 and Kolmogorov's 0-1 law Theorem 6.1.1. A proof can also be based on the forward martingale representation discussed in Sect. 8.1.

Theorem 6.3.1 (U-statistics SLLN) *Let $\{X_i\}$ be a sequence of iid random variables on $(\Omega, \mathcal{A}, P)$. Let U_n be the U-statistic with kernel h, and $E(|h(X_1, \ldots, X_m)|) < \infty$. Then,*

$$\lim_{n \to \infty} U_n = E(h(X_1, \ldots, X_m)), \quad \text{almost surely, and in } L^1.$$

◆

To prove the theorem, we need a few definitions which shall be useful to us later also. Fix $n \in \mathbb{N}$. Let $\text{perm}(1, \ldots, n)$ be the set of permutations of $(1, \ldots, n)$. For $\pi_n \in \text{perm}(1, \ldots, n)$, let $f_{\pi_n} : \mathbb{R}^\mathbb{N} \to \mathbb{R}^\mathbb{N}$ be

$$f_{\pi_n}(x) := (x_{\pi(1)}, \ldots, x_{\pi(n)}, x_{n+1}, \ldots).$$

Thus f_{π_n} applies π_n to the first n co-ordinates of x, and leaves the other co-ordinates intact. Define the σ-fields $\mathcal{P}_\pi, \mathcal{P}_n, \mathcal{P}$ and $\mathcal{G}_n$ as

$$\mathcal{P}_{\pi_n} := \{B \in \mathcal{B}(\mathbb{R}^\mathbb{N}) : f_{\pi_n}(B) = B\} \quad (6.6)$$

$$\mathcal{P}_n := \bigcap_{\pi_n \in \text{perm}(1, \ldots, n)} \mathcal{P}_{\pi_n}, \quad (6.7)$$

$$\mathcal{P} := \bigcap_{n=1}^{\infty} \mathcal{P}_n, \text{ and} \quad (6.8)$$

$$\mathcal{G}_n := \{A \in \mathcal{A} :$$
$$A = \{(X_1, X_2, \ldots) \in B\} \text{ for some } B \in \mathcal{P}_n\}. \quad (6.9)$$

Proof of Theorem 6.3.1 Let $\mathcal{P}_n$ be as in (6.7). It immediately follows that $\mathcal{P}_n \supset \mathcal{P}_{n+1}$, and hence $\mathcal{G}_n \supset \mathcal{G}_{n+1}$. Clearly U_n is $\mathcal{G}_n$-measurable. We claim that $(U_n, \mathcal{G}_n)_{n \geq 1}$ is a reverse martingale.

Since $\{X_i\}$ is an iid sequence, for distinct $i_1, \ldots, i_m \in \{1, \ldots, n\}, n \geq m$,

$$E\big(h(X_1, \ldots, X_m)|\mathcal{G}_n\big) = E\big(h(X_{i_1}, \ldots, X_{i_m})|\mathcal{G}_n\big).$$

Averaging over all choices of distinct $i_1, \ldots, i_m \in \{1, \ldots, n\}$, we get

$$E\big(h(X_1, \ldots, X_m)|\mathcal{G}_n\big) = E\left(\frac{1}{m!}\binom{n}{m}^{-1} \sum_{\substack{i_1, \ldots, i_m \in \{1, \ldots, n\} \\ i_1, \ldots, i_m \text{ distinct}}} h(X_{i_1}, \ldots, X_{i_m}) \Big| \mathcal{G}_n\right)$$

$$= E(U_n|\mathcal{G}_n)$$
$$= U_n \quad (\text{since } U_n \text{ is } \mathcal{G}_n\text{-measurable}).$$

Hence $(U_n, \mathcal{G}_n)_{n\geq 1}$ is a reverse martingale. By the Reverse martingale convergence Theorem 5.4.1, there exists U_∞ such that as $n \to \infty$,

$$U_n \to U_\infty \text{ almost surely, and in } L^1. \tag{6.10}$$

It is now enough to show that U_∞ is a constant almost surely.

Let $\mathcal{A}_i$ be the completion of $\sigma(X_i)$, $i \geq 1$. By Exercise 1.12.4, $\{\mathcal{A}_i\}_{i\geq 1}$ are independent.

Fix $k \geq 2$ and $n \geq k + m$. Define

$$A_n := \big\{\{i_1, \ldots, i_m\} : \{i_j\} \text{ distinct}, \{i_1, \ldots, i_m\} \cap \{1, \ldots, k-1\} \neq \emptyset\big\}.$$

Then we can write

$$U_n = \frac{1}{m!\binom{n}{m}} \sum_{\{i_1, \ldots, i_m\} \in A_n^c} h(X_{i_1}, \ldots, X_{i_m}) + \frac{1}{m!\binom{n}{m}} \sum_{\{i_1, \ldots, i_m\} \in A_n} h(X_{i_1}, \ldots, X_{i_m})$$

$$=: V_n + W_n.$$

For the second term, note that

$$E(|W_n|) \leq \frac{1}{m!}\binom{n}{m}^{-1} (\#A_n) E\big(|h(X_1, \ldots, X_m)|\big)$$

$$= \frac{1}{m!}\binom{n}{m}^{-1} \left(\binom{n}{m} - \binom{n-k+1}{m}\right) E\big(|h(X_1, \ldots, X_m)|\big)$$

$$\to 0 \text{ as } n \to \infty.$$

Thus, $W_n \xrightarrow{P} 0$, and hence recalling (6.10), it follows that

$$V_n \xrightarrow{P} U_\infty \text{ as } n \to \infty.$$

By Exercise 1.13.7, there is a sub-sequence of $\{V_n\}$ which converges to U_∞ almost surely. We continue to denote this sub-sequence by $\{V_n\}$.

Observe that each V_n, for $n \geq k + m$, is $\bigvee_{i=k}^\infty \mathcal{A}_i$-measurable. Since each $\mathcal{A}_i$ is complete, by Exercise 1.3.2, $\bigvee_{i=k}^\infty \mathcal{A}_i$ is also complete. It follows that U_∞ is also measurable with respect to $\bigvee_{i=k}^\infty \mathcal{A}_i$, for all k. Hence, U_∞ is measurable with respect to $\bigcap_{k=1}^\infty \bigvee_{i=k}^\infty \mathcal{A}_i$.

By Kolmogorov's 0-1 law Theorem 6.1.1, this σ-field is trivial, and hence U_∞ is a constant almost surely. Hence $U_\infty = \mathrm{E}\left(h(X_1, \ldots, X_m)\right)$, which in view of (6.10) completes the proof. $\square$

6.4 Hewitt–Savage 0-1 Law

If the independent σ-fields in Kolmogorov's 0-1 law Theorem 6.1.1 are generated by iid random variables, then a σ-field larger than the tail σ-field becomes trivial. This is the result of Hewitt and Savage (1955).

Definition 6.4.1 (*Tail invariant σ-field*) Suppose that $\{X_i\}$ is a sequence of random variables on $(\Omega, \mathcal{A}, P)$. Define the **tail invariant σ-field** with respect to $\{X_i\}$ as

$$\mathcal{T}_I := \{A \in \mathcal{A} : A = \{\omega : (X_1(\omega), X_2(\omega), \ldots) \in B \text{ for some } B \in \mathcal{P}\}\},$$

where $\mathcal{P}$ is as in (6.8). ◇

Note that the sets in $\mathcal{T}_I$ are invariant with respect to permutation of any finite number of X_i in the sequence $\{X_i\}$.

Theorem 6.4.1 (Hewitt–Savage 0-1 law) *Let $\{X_i\}$ be iid random variables. Then $\mathcal{T}_I$ is trivial.* ◆

Remark 6.4.1 Before we prove the theorem, we note that the Hewitt–Savage 0-1 law implies Kolmogorov's 0-1 law in the iid case. This will follow as soon as we realize that $\mathcal{T}_I$ is larger than $\mathcal{T}$.

To see this, suppose that $\{X_i\}$ are iid random variables defined on $(\Omega, \mathcal{A}, P)$. Suppose $A \in \sigma(X_{n+1}, X_{n+2}, \ldots)$ for some n. Then it follows that for every permutation π_n of $(1, \ldots, n)$, there is a $B \in \mathcal{P}_{\pi_n}$ such that $A = \{\omega \in \Omega : (X_1(\omega), X_2(\omega), \ldots) \in B\}$. As a consequence,

$$\mathcal{T} = \bigcap_{n=1}^\infty \sigma(X_n, X_{n+1}, \ldots)$$
$$\subset \{A \in \mathcal{A} : A = \{(X_1, X_2, \ldots) \in B\} \text{ for some } B \in \mathcal{P}\}$$
$$= \mathcal{T}_I.$$

●

Proof of Theorem 6.4.1 Fix $B \in \mathcal{P}$, and define $W := \mathbf{1}_{\{(X_1, X_2, \ldots) \in B\}}$. In order to complete the proof, it suffices to show that

$$\mathrm{E}(W^2) = [\mathrm{E}(W)]^2. \tag{6.11}$$

Let $\mathcal{A}_n := \sigma(X_1, \ldots, X_n)$, $n \geq 1$. Note that W is $\bigvee_{n=1}^{\infty} \mathcal{A}_n$-measurable and $0 \leq W \leq 1$.

Hence, for any $n \geq 1$, there exists a measurable function $g : \mathbb{R}^n \to [0, 1]$ (depending on n) such that

$$Y := \mathrm{E}(W | \mathcal{A}_n) = g(X_1, \ldots, X_n).$$

Fix $\varepsilon > 0$. By Lévy's upward Theorem 5.2.4, $\mathrm{E}(W|\mathcal{A}_n) \to W$ almost surely and in L^1 as $n \to \infty$.

Hence there exists $n \geq 1$ such that

$$\mathrm{E}\left(|\mathbf{1}_{\{(X_1, X_2, \ldots) \in B\}} - g(X_1, \ldots, X_n)|\right) \leq \varepsilon. \tag{6.12}$$

Since $\{X_i\}$ are iid,

$$(X_1, X_2, \ldots) \stackrel{d}{=} (X_{n+1}, \ldots, X_{2n}, X_1, \ldots, X_n, X_{2n+1}, X_{2n+2}, \ldots).$$

Hence from (6.12) it follows that

$$\mathrm{E}\left(|\mathbf{1}_{\{(X_{n+1}, \ldots, X_{2n}, X_1, \ldots, X_n, X_{2n+1}, X_{2n+2}, \ldots) \in B\}} - g(X_{n+1}, \ldots, X_{2n})|\right) \leq \varepsilon. \tag{6.13}$$

Since $B \in \mathcal{P}$, W is invariant with respect to finite permutations, and hence

$$W = \mathbf{1}_{\{(X_{n+1}, \ldots, X_{2n}, X_1, \ldots, X_n, X_{2n+1}, X_{2n+2}, \ldots) \in B\}}. \tag{6.14}$$

Let $Z := g(X_{n+1}, \ldots, X_{2n})$. Then (6.13) and (6.14) imply that

$$\mathrm{E}(|W - Z|) \leq \varepsilon. \tag{6.15}$$

The variables Y and Z are iid, $0 \leq Y, Z \leq 1$, and $\mathrm{E}(Y) = \mathrm{E}(Z) = \mathrm{E}(W)$. Hence,

$$\begin{aligned}|\mathrm{E}(W^2) - [\mathrm{E}(W)]^2| &= |\mathrm{E}(W^2) - \mathrm{E}(YZ)| \tag{6.16}\\ &\leq \mathrm{E}(|W^2 - WY|) + \mathrm{E}(|WY - YZ|)\\ &\leq \mathrm{E}(|W - Y|) + \mathrm{E}(|W - Z|)\\ &\leq 2\varepsilon, \tag{6.17}\end{aligned}$$

where (6.17) follows from (6.12) and (6.15). Since ε is arbitrary, this confirms (6.11) and hence the proof is complete. $\square$

6.5 Strong Law for Exchangeable Sequences

Definition 6.5.1 (*Exchangeable sequence*) A sequence of random variables $\{X_i\}$ is said to be **exchangeable** if, for all $n \geq 1$ and every permutation π of $(1, \ldots, n)$,

$$(X_1, \ldots, X_n) \stackrel{d}{=} (X_{\pi(1)}, \ldots, X_{\pi(n)}).$$

◇

For more information on exchangeable sequences, see Aldous (1985). In Sect. 6.6, we shall present a beautiful representation theorem for such sequences. In this section we state and prove a SLLN for exchangeable sequences. Unlike in the iid case, the tail σ-field $\mathcal{G}_\infty$ need not be trivial, and hence the limit need not be a constant almost surely.

Example 6.5.1 Suppose $Z_0, Z_1, \ldots$ are iid standard Gaussian random variables. Define
$$X_n := \sqrt{\rho} Z_0 + \sqrt{1-\rho}\, Z_n,\ n \geq 1,\ 0 \leq \rho < 1.$$

Then, $\{X_i\}$ is also a Gaussian sequence. In particular, for every n, $(X_1, \ldots, X_n)$ follows a multivariate Gaussian distribution, and for all $1 \leq i, j \leq n$, $\mathrm{E}(X_i) = 0$. In addition,

$$\mathrm{Cov}(X_i, X_j) = \begin{cases} \rho & \text{if } i \neq j, \\ 1 & \text{if } i = j. \end{cases}$$

It is then easy to see that the sequence $\{X_i\}$ is exchangeable. ▲

Exercise 6.5.1 Suppose $\{X_i\}_{1 \leq i \leq n}$ is jointly Gaussian for some $n \geq 1$. Give a necessary and sufficient condition for this finite sequence to be exchangeable. When is this finite sequence extendible to an infinite sequence of exchangeable random variables?

Example 6.5.2 Let Y be a random variable, taking values in the interval $[0, 1]$. Conditional on Y, let $\{X_i\}$ be iid Bernoulli random variables with parameter Y. That is, $P\{X_1 = 1|Y\} = 1 - P\{X_1 = 0|Y\} = Y$ almost surely. Then $\{X_i\}$ is exchangeable, and is iid if and only if Y is a constant almost surely. We shall see below in Sect. 6.6 that every $\{0, 1\}$-valued exchangeable sequence $\{X_i\}$ arises in this way. ▲

Theorem 6.5.1 (SLLN for exchangeable sequences) *Let $\{X_i\}$ be an exchangeable sequence of integrable random variables. Then there exists an integrable random variable Y such that as $n \to \infty$,*

$$\frac{1}{n} \sum_{i=1}^{n} X_i \to Y \text{ almost surely, and in } L^1.$$

♦

Proof Let $S_n := \sum_{i=1}^n X_i$, $n \geq 1$. Define the σ-fields

$$\mathcal{G}_{mn} := \sigma(S_m, S_{m+1}, \ldots, S_n),\ 1 \leq m \leq n < \infty,$$
$$\mathcal{G}_{m\infty} := \sigma(S_m, S_{m+1}, \ldots),\ m \in \mathbb{N},$$
$$\mathcal{G}_\infty := \bigcap_{m=1}^\infty \mathcal{G}_{m\infty}.$$

Then $\mathcal{G}_\infty$ is the tail σ-field of $\{\sigma(S_n)\}$. Fix $2 \leq m \leq n$. We first show that

$$\mathrm{E}(X_1|\mathcal{G}_{mn}) = m^{-1} S_m. \tag{6.18}$$

First note that the left side of (6.18) is $\mathcal{G}_{mn}$-measurable. Hence there exists a Borel measurable function $f : \mathbb{R}^{n-m+1} \to \mathbb{R}$ such that

$$\mathrm{E}(X_1|\mathcal{G}_{mn}) = f(S_m, \ldots, S_n). \tag{6.19}$$

On the other hand, for any $i \in \{1, \ldots, m\}$,

$$(X_1, X_2, \ldots, X_n) \stackrel{d}{=} (X_i, X_1, X_2, \ldots, X_{i-1}, X_{i+1}, \ldots, X_n).$$

Hence, for any $i \in \{1, \ldots, m\}$,

$$(X_1, S_m, \ldots, S_n) \stackrel{d}{=} (X_i, S_m, \ldots, S_n).$$

Therefore, using (6.19),

$$\mathrm{E}(X_i|\mathcal{G}_{mn}) = f(S_m, \ldots, S_n),\ 1 \leq i \leq m,$$
$$mf(S_m, \ldots, S_n) = \sum_{i=1}^m \mathrm{E}(X_i|\mathcal{G}_{mn}) = \mathrm{E}(S_m|\mathcal{G}_{mn}) = S_m.$$

This establishes (6.18). Now, for any m, $\mathcal{G}_{m\infty} = \bigvee_{n=1}^\infty \mathcal{G}_{mn}$. Therefore, by Lévy's upward Theorem 5.2.4, and (6.18),

$$\mathrm{E}(X_1|\mathcal{G}_{m\infty}) = \lim_{n\to\infty} \mathrm{E}(X_1|\mathcal{G}_{mn}) = m^{-1} S_m.$$

Observe that $\{\mathcal{G}_{m\infty}\}_{m\in\mathbb{N}}$ is a reverse filtration. Therefor, by the Reverse Martingale convergence Theorem 5.4.1,

$$\lim_{m\to\infty} m^{-1} S_m \to \mathrm{E}(X_1|\mathcal{G}_\infty)\ \text{almost surely, and in}\ L^1.$$

So the theorem is proved with $Y = \mathrm{E}(X_1|\mathcal{G}_\infty)$. □

6.6 de Finetti's Theorem

de Finetti's theorem (de Finetti 1972) says that any sequence of exchangeable random variables is conditionally iid. We first prove it for the special case when the random variables take two values. In particular, this establishes the claim made in Example 6.5.2. The proof involves a non-trivial application of Theorem 6.5.1. For an alternate proof in this special case that uses the moment method, see Kirsch (2019).

Theorem 6.6.1 (de Finetti's Theorem: the Bernoulli case) *Let $\{X_n\}$ be an exchangeable sequence of 0-1 valued random variables. Then there exists a random variable Y, $0 \leq Y \leq 1$ such that for all $n \geq 1$,*

$$P(X_1 = u_1, \ldots, X_n = u_n | \sigma(Y)) = Y^{\sum_{i=1}^n u_i}(1-Y)^{n - \sum_{i=1}^n u_i},$$

for all $u_1, \ldots, u_n \in \{0, 1\}$, with the convention that $0^0 = 1$. So, given Y, $\{X_i\}$ are iid Bernoulli variables with success probability Y. ◆

Proof For $n \in \mathbb{N}$, and $1 \leq m \leq n \leq \infty$, let S_n and $\mathcal{G}_{mn}$ be as defined in the proof of Theorem 6.5.1. Fix $1 \leq m \leq n < \infty$. Fix $v_m, \ldots, v_n$ such that

$$D_{mn} := P(S_m = v_m, \ldots, S_n = v_n) > 0.$$

Fix $u_1, \ldots, u_m \in \{0, 1\}$ such that $u_1 + \cdots + u_m = v_m$, and observe that

$$P(X_1 = u_1, \ldots, X_m = u_m | S_m = v_m, \ldots, S_n = v_n)$$
$$= \frac{P(X_i = u_i, 1 \leq i \leq m; X_{j+1} = v_{j+1} - v_j, j \leq m \leq n-1)}{D_{mn}}.$$

Clearly, there exist $1 \leq i_1 < \cdots < i_{v_m} \leq m$ so that $u_{i_1} = \cdots = u_{i_{v_m}} = 1$ and $u_j = 0$ for $j \in \{1, \ldots, m\} \setminus \{i_1, \ldots, i_{v_m}\}$.

Therefore,

$$P(X_i = u_i, 1 \leq i \leq m; X_{j+1} = v_{j+1} - v_j, j \leq m \leq n-1)$$
$$= P(X_j = 1, j \in \{i_1, \ldots, i_{v_m}\}; X_j = 0, j \in \{1, \ldots, m\} \setminus \{i_1, \ldots, i_{v_m}\};$$
$$X_j = v_j - v_{j-1} \text{ for } j = m+1, \ldots, n)$$
$$= P(X_j = 1 \text{ for } j \in \{1, \ldots, v_m\}, X_j = 0 \text{ for } j \in \{v_m + 1, \ldots, m\},$$
$$X_j = v_j - v_{j-1} \text{ for } j = m+1, \ldots, n) \text{ (by exchangeability)}.$$

That is, for all $u_1, \ldots, u_m \in \{0, 1\}$ with $u_1 + \cdots + u_m = v_m$,

$$P(X_1 = u_1, \ldots, X_m = u_m | S_m = v_m, \ldots, S_n = v_n)$$

depends only on $v_m, \ldots, v_n$.

As there are $\binom{m}{v_m}$ choices for $u_1, \ldots, u_m$, it follows that

$$P(X_1 = u_1, \ldots, X_m = u_m | S_m = v_m, \ldots, S_n = v_n) = \binom{m}{v_m}^{-1},$$

for all $u_1, \ldots, u_m \in \{0, 1\}$ with $u_1 + \cdots + u_m = v_m$.

That is, for all $1 \leq m \leq n < \infty$,

$$P(X_1 = u_1, \ldots, X_m = u_m | \mathcal{G}_{mn}) = \binom{m}{S_m}^{-1} \mathbf{1}_{\{S_m = \sum_{i=1}^m u_i\}}.$$

Letting $n \to \infty$, it follows that for all $1 \leq m < \infty$,

$$P(X_1 = u_1, \ldots, X_m = u_m | \mathcal{G}_{m\infty}) = \binom{m}{m}^{-1} \mathbf{1}_{\{S_m = \sum_{i=1}^m u_i\}}. \quad (6.20)$$

Fix once again $1 \leq m \leq n < \infty$, and $u_1, \ldots, u_m \in \{0, 1\}$. Define

$$j := \sum_{i=1}^m u_i, \quad k := m - j,$$

and note that

$$P(X_1 = u_1, \ldots, X_m = u_m | \mathcal{G}_{n\infty})$$
$$= \frac{\binom{n-m}{S_n-j}}{\binom{n}{S_n}} \mathbf{1}_{\{j \leq S_n \leq n-k\}} \quad \text{(by (6.20) and exchangeability)}$$
$$= \mathbf{1}_{\{j \leq S_n \leq n-k\}} \frac{\left(\prod_{i=0}^{j-1} S_{n-i}\right)\left(\prod_{i=0}^{k-1}(n - S_n - i)\right)}{n(n-1)\cdots(n-m+1)}$$
$$= \mathbf{1}_{\{j \leq S_n \leq n-k\}} \frac{S_n}{n} \cdots \frac{S_n - j + 1}{n - j + 1} \frac{n - S_n}{n - j} \cdots \frac{n - S_n - k + 1}{n - m + 1}$$
$$=: \mathbf{1}_{\{j \leq S_n \leq n-k\}} Y_{mn} \quad \text{(say)}.$$

Fixing m and letting $n \to \infty$, it is easy to see by Theorem 6.5.1 that

$$Y_{mn} \to Y^j (1 - Y)^k, \text{ almost surely},$$

where Y is the almost sure limit of S_n/n. Thus, it is immediate that on $\{0 < Y < 1\}$,

$$\lim_{n \to \infty} P(X_1 = u_1, \ldots, X_m = u_m | \mathcal{G}_{n\infty}) = Y^j (1 - Y)^k, \text{ almost surely}. \quad (6.21)$$

Now we consider the case $Y = 0$. Fix ω such that $\frac{1}{n} S_n(\omega) \to Y(\omega) = 0$. If $j \geq 1$, then it is easy to see that

6.6 de Finetti's Theorem

$$P(X_1 = u_1, \ldots, X_m = u_m | \mathcal{G}_{n\infty})(\omega) \leq \frac{1}{n} S_n(\omega) \to 0.$$

If $j = 0$, then

$$P(X_1 = u_1, \ldots, X_m = u_m | \mathcal{G}_{n\infty})(\omega) = \mathbf{1}_{\{S_n(\omega) \leq n-m\}} \prod_{i=0}^{m-1} \frac{n - S_n(\omega) - i}{n - m + 1}$$

$$\to 1, \text{ as } n \to \infty.$$

Thus, recalling the convention that $0^0 = 1$, (6.21) holds on $\{Y = 0\}$. The convergence on $\{Y = 1\}$ can be established similarly.

By the Reverse martingale convergence Theorem 5.4.1, as $n \to \infty$,

$$P(X_1 = u_1, \ldots, X_m = u_m | \mathcal{G}_{n\infty}) \to P(X_1 = u_1, \ldots, X_m = u_m | \mathcal{G}_\infty) \text{ a.s.}$$

From (6.21) it follows that

$$P(X_1 = u_1, \ldots, X_m = u_m | \mathcal{G}_\infty) = Y^{\sum_{i=1}^m u_i} (1 - Y)^{m - \sum_{i=1}^m u_i}.$$

The proof can now be completed by taking conditional expectation with respect to $\sigma(Y)$ which is a sub-σ-field of $\mathcal{G}_\infty$. □

The more general form of de-Finetti's theorem does not restrict to $\{0, 1\}$ random variables. We need some definitions and notation. Note that the definitions given below make sense due to the existence of RCD; see Theorem 3.2.1. We shall not mention this again.

Definition 6.6.1 (*Conditional distribution function*) Let $(\Omega, \mathcal{A}, P)$ be a probability space, X be a random variable, and $\mathcal{G}$ be a sub-σ-field of $\mathcal{A}$. Then the (regular) **conditional distribution function** of X given $\mathcal{G}$ is defined as

$$F(x|\mathcal{G}) := P(X \leq x | \mathcal{G}), \ x \in \mathbb{R}.$$

This extends to joint (regular) conditional distribution function of a sequence of random variables in a natural way. Clearly, $F(x|\mathcal{G})$ is random and satisfies all the conditions of a CDF. ◇

Definition 6.6.2 (*Conditional independence*) Suppose $(\Omega, \mathcal{A}, P)$ is a probability space. Suppose $\mathcal{G} \subset \mathcal{A}$ is a σ-field. Any sequence of random variables $\{X_n\}$ is said to be **conditionally independent** given $\mathcal{G}$, if for all $k \geq 1$ and all Borel sets $\{A_i\}_{1 \leq i \leq k}$,

$$P(X_i \in A_i, 1 \leq i \leq k | \mathcal{G}) = \prod_{i=1}^k P(X_i \in A_i | \mathcal{G}), \text{ almost surely.} \quad (6.22)$$

This definition can be extended in the obvious way to conditional independence between random variables and σ-fields or between σ-fields, conditioned on a given σ-field. ◇

Note that (6.22) can be expressed in the equivalent way:

$$P(X_i \leq x_i, 1 \leq i \leq k | \mathcal{G}) = \prod_{i=1}^{k} P(X_i \leq x_i | \mathcal{G}), \text{ for all } x_i \in \mathbb{R}.$$

Theorem 6.6.2 (de Finetti's theorem) *Suppose $\{X_n\}$ is an exchangeable sequence on $(\Omega, \mathcal{A}, P)$. Then there is a σ-field $\mathcal{G} \subset \mathcal{A}$ such that $\{X_n\}$ is conditionally iid given $\mathcal{G}$.* ♦

Proof Let $\mathcal{P}_\pi$, $\mathcal{P}_n$ and $\mathcal{G}_n$ be as defined in (6.6), (6.7) and (6.9). Then $\mathcal{G}_n \supset \mathcal{G}_{n+1}$ for $n \geq 1$. Let $\mathcal{G}_\infty := \bigvee_{n=1}^{\infty} \mathcal{G}_n$. Fix $x \in \mathbb{R}$, $n \geq 1$, and let $Y_n := \mathbf{1}_{\{X_n \leq x\}}$, $S_n := Y_1 + \cdots + Y_n$. Using exchangeability, it is easy to check that $(n^{-1} S_n, \mathcal{G}_n)_{n \geq 1}$ is a reverse martingale. Hence,

$$n^{-1} \sum_{i=1}^{n} \mathbf{1}_{\{X_i \leq x\}} = \mathrm{E}\big(\mathbf{1}_{\{X_1 \leq x\}} | \mathcal{G}_n\big)$$

$$\to \mathrm{E}\big(\mathbf{1}_{\{X_1 \leq x\}} | \mathcal{G}_\infty\big) \qquad (6.23)$$

$$= F(x | \mathcal{G}_\infty) \text{ almost surely,} \qquad (6.24)$$

where $F(x | \mathcal{G}_\infty)$ is the conditional distribution of X_1 given $\mathcal{G}_\infty$. Now fix k and take $n \geq k$. By exchangeability of $\{X_i\}$, and the nature of $\{\mathcal{G}_i\}$,

$$\mathrm{E}\big(\prod_{i=1}^{k} \mathbf{1}_{\{X_i \leq x_i\}} | \mathcal{G}_n\big) = [\prod_{i=0}^{k-1}(n-i)]^{-1} \sum_{\substack{1 \leq j_i \leq n, 1 \leq i \leq k \\ j_1, \ldots, j_k \text{ distinct}}} \prod_{i=1}^{k} \mathbf{1}_{\{X_{j_i} \leq x_i\}}$$

$$= p_{nk} \text{ (say).} \qquad (6.25)$$

Using the Reverse martingale convergence Theorem 5.4.1, left side of (6.25) converges to

$$\mathrm{E}\big(\prod_{i=1}^{k} \mathbf{1}_{\{X_i \leq x_i\}} | \mathcal{G}_\infty\big) = P\big(X_i \leq x_i, 1 \leq i \leq k | \mathcal{G}_\infty\big).$$

On the other hand, for the right side of (6.25), if we introduce the equality terms, then the contribution of these terms would be almost surely negligible as $n \to \infty$. Hence for every $k \geq 1$,

$$\lim_{n \to \infty} p_{n,k} = \lim_{n \to \infty} n^{-k} \sum_{1 \leq j_i \leq n, 1 \leq i \leq k} \prod_{i=1}^{k} \mathbf{1}_{\{X_{j_i} \leq x_i\}}$$

$$= \lim_{n\to\infty} \prod_{i=1}^{k} \left[n^{-1} \sum_{1\leq j_i \leq n} \mathbf{1}_{\{X_{j_i} \leq x_i\}} \right] = \prod_{i=1}^{k} F(x_i | \mathcal{G}_\infty).$$

This completes the proof. □

6.7 Kakutani's Theorem for Martingales

We begin this section with a simple exercise.

Exercise 6.7.1 (a) Suppose X is a non-negative random variable with $E(X) = 1$. Show that $0 < E\left(\sqrt{X}\right) \leq 1$ for all n.

(b) Let $\{b_n\}$ be any sequence of real numbers where $0 < b_n \leq 1$ for all $n \geq 1$. Show that
$$\prod_{n\geq 1} b_n > 0 \text{ if and only if } \sum_{n\geq 1} (1 - b_n) < \infty.$$

We recall the concept of product martingale from Exercise 4.1.6.

Definition 6.7.1 Suppose $\{Y_i\}$ are independent, non-negative random variables, and $E(Y_i) = 1$ for all $i \geq 1$. Let $X_n = \prod_{i=1}^n Y_i$. Then $(X_n, \mathcal{G}_n := \sigma(Y_1, \ldots, Y_n))$ is called a **product martingale**. ◇

Let $(X_n, \mathcal{G}_n)$ be a product martingale. As X_n is non-negative, and for all n, $E(X_n) = 1$, by the Martingale convergence theorem, $X_n \to X_\infty \geq 0$ almost surely as $n \to \infty$, and X_∞ is finite a.s.

Theorem 6.7.1 (Kakutani's theorem) *Let* $X_n = \prod_{i=1}^n Y_i$ *where* $\{Y_i\}$ *are independent, non-negative random variables, and* $E(Y_i) = 1$ *for all* $i \geq 1$. *Let* $X_\infty = \lim X_n$ *and* $a_n = E\left(\sqrt{Y_n}\right)$. *Then the following are equivalent.*

(i) $P(X_\infty > 0) > 0$;

(ii) $\prod_{n\geq 1} a_n > 0$;

(iii) $(X_n, n \geq 1)$ *is uniformly integrable*;

(iv) $E(X_\infty) = 1$. ◆

Proof (i) $\Longrightarrow$ (ii): Note that $0 < a_n < 1$ for all n. Let
$$M_n = \prod_{i=1}^{n} \frac{\sqrt{Y_i}}{a_i}$$
$$= \frac{1}{\prod_{i=1}^n a_i} \sqrt{X_n}.$$

Now, M_n is a non-negative product martingale with $E(M_n) = 1$ for all $n \geq 1$. By the martingale convergence theorem, $M_n \to M_\infty \geq 0$ a.s. where M_∞ is finite almost surely.

Suppose now that $\prod_{n\geq 1} a_n = 0$. This implies that $\prod_{n\geq 1} a_n^2 = 0$. Hence $X_n = M_n^2 \prod_{i=1}^n a_i^2 \to 0 = X_\infty$ almost surely, which would be a contradiction. Hence $\prod_{n\geq 1} a_n > 0$.

(ii) $\Longrightarrow$ (iii): Assume that $\prod_{n\geq 1} a_n > 0$. Recall that $0 < a_n \leq 1$ for all n. Hence

$$E\left(M_n^2\right) = E\left(\frac{X_n}{\prod_{i=1}^n a_i^2}\right) = \frac{1}{\prod_{i=1}^n a_i^2} \leq \frac{1}{\prod_{i\geq 1} a_i^2} < \infty.$$

Therefore $\{M_n\}$ is bounded in L^2. Let

$$M_n^* = \max_{1 \leq i \leq n} M_i,$$

$$M_\infty^* = \sup_{n\geq 1} M_n^* = \sup_{n\geq 1} \frac{1}{\prod_{i=1}^n a_i} \sqrt{X_n}.$$

Note that M_n^* increases to M_∞^*. Using the inequality proved in Theorem 5.3.1 (see (5.7) for $p = 2$) we have

$$E\left((M_n^*)^2\right) \leq \left(\frac{2}{2-1}\right)^2 E\left(M_n^2\right) \leq \frac{4}{\prod_{i\geq 1} a_i^2},$$

Therefore

$$M_\infty^* = \sup_{n\geq 0} \frac{1}{\prod_{i=1}^n a_i} \sqrt{X_n}$$

and

$$E\left((M_\infty^*)^2\right) = \sup_{n\geq 0} E\left((M_n^*)^2\right) < \infty.$$

On the other hand,

$$\left(M_\infty^*\right)^2 = \sup_{n\geq 0} \frac{X_n}{\prod_{i=1}^n a_i^2} \quad \text{and thus} \quad \sup_{n\geq 0} X_n \leq \left(M_\infty^*\right)^2.$$

As $(X_n, n \geq 1)$ is dominated by the integrable r.v. $\left(M_\infty^*\right)^2$, it is UI.

(iii) $\Longrightarrow$ (iv): If $(X_n, n \geq 1)$ is uniformly integrable, then $X_n \to X_\infty$ a.s. and in L^1 as $n \to \infty$. We have $1 = E(X_n) \to E(X_\infty)$.

(iv) $\Longrightarrow$ (i): If $E(X_\infty) = 1$ then $P(X_\infty > 0) > 0$ since $X_\infty \geq 0$ a.s. $\square$

Remark 6.7.1 Consider the case where $P(Y_n > 0) = 1$ for all n. Then $P(X_n > 0) = 1$ for all n as well. Therefore $A := \{X_\infty > 0\}$ is a tail event. By Kolmogorov's 0-1 law, $P(A) \in \{0, 1\}$. Hence $P(X_\infty > 0) > 0$ if and only if $P(X_\infty > 0) = 1$. ●

6.8 Likelihood Ratio

The likelihood ratio is a fundamental object in statistics. In this section we shall study this quantity via martingales.

Suppose $(\Omega, \mathcal{A})$ is a measurable space with two probability measures P and Q. Suppose $\{X_i\}$ is a sequence of random variables. For each n, consider the measures $Q_{X_1,\ldots,X_n}$ and $P_{X_1,\ldots,X_n}$ induced on $\mathbb{R}^n$ by $(X_1, \ldots, X_n)$ under Q and P respectively. Suppose they have densities q_n and p_n with respect to the Lebesgue measure λ_n on $\mathbb{R}^n$. Suppose that

$$p_n(\cdot) = 0 \text{ implies } q_n(\cdot) = 0. \tag{6.26}$$

Definition 6.8.1 The **likelihood ratio** is defined for all $n \geq 1$ as

$$L_n := L_n(X_1, \ldots, X_n) = \begin{cases} \frac{q_n(X_1,\ldots,X_n)}{p_n(X_1,\ldots,X_n)} & \text{if } p_n(X_1, \ldots, X_n) \neq 0, \\ 0 & \text{if } p_n(X_1, \ldots, X_n) = 0. \end{cases}$$

◇

Theorem 6.8.1 *Let P and Q be probability measures on the measurable space $(\Omega, \mathcal{A})$, and $\{X_i\}$ be random variables on this space.*
(a) Let Q_n and P_n be the restrictions of Q and P on $\mathcal{G}_n = \sigma(X_1, \ldots, X_n)$ with densities q_n and p_n which satisfy (6.26). Then $Q_n \ll P_n$ and

$$\frac{dQ_n}{dP_n} = L_n \text{ almost surely } P,$$

where L_n is the likelihood ratio.
(b) $(L_n, \mathcal{G}_n = \sigma(X_0, \ldots, X_n))_{n \geq 1}$ is a martingale on $(\Omega, \mathcal{A}, P)$. ♦

Proof (a) Let λ_n denote the Lebesgue measure on $\mathbb{R}^n$. Any $A \in \mathcal{G}_n$ is of the form

$$A = \{\omega \in \Omega : (X_1(\omega), \ldots, X_n(\omega)) \in B_n\} \text{ for some Borel set } B_n \subset \mathbb{R}^n.$$

Writing $x = (x_1, \ldots x_n)$,

$$Q_n(A) = Q_n\{(X_1, \ldots, X_n) \in B_n\} = \int_{B_n} q_n(x)\lambda_n(dx)$$

$$= \int_{B_n} \frac{q_n(x)}{p_n(x)} p_n(x)\lambda_n(dx) \text{ (by (6.26))}$$

$$= E_{P_n}\left(L_n \mathbf{1}_{(X_1,\ldots,X_n) \in B_n}\right).$$

(b) Note that the conditional probability distribution of X_{n+1} given $(X_1, \ldots, X_n)$ has density $\frac{p_{n+1}(X_1,\ldots,X_n,x)}{p_n(X_1,\ldots,X_n)}$ (the set where the denominator is 0 does not count). Hence

$$\mathrm{E}_P(L_{n+1}(X_1,\ldots,X_{n+1})|\mathcal{G}_n)$$
$$= \int \frac{q_{n+1}(X_1,\ldots,X_n,x)}{p_{n+1}(X_1,\ldots,X_n,x)} \frac{p_{n+1}(X_1,\ldots X_n,x)}{p_n(X_1,\ldots X_n)} \lambda(dx)$$
$$= \int \frac{q_{n+1}(X_1,\ldots,X_n,x)}{p_n(X_1,\ldots X_n)} \lambda(dx)$$
$$= \frac{q_n(X_1,\ldots,X_n)}{p_n(X_1,\ldots X_n)} = L_n.$$

□

Now suppose that $\{X_n\}$ are independent with densities $\{f_n\}$ and $\{g_n\}$ respectively under P and Q where for each n,

$$f_n = 0 \text{ implies } g_n = 0. \tag{6.27}$$

Let

$$Y_n = \frac{g_n(X_n)}{f_n(X_n)} \text{ and } L_n = \prod_{i=1}^{n} Y_i.$$

Then $(L_n, \mathcal{G}_n)$ is a product martingale with respect to P, and hence $L_n \to L_\infty = \prod_{n=1}^{\infty} \frac{g_n(X_n)}{f_n(X_n)}$ almost surely with respect to P.

Theorem 6.8.2 *Suppose for two probability measure P and Q, on $(\Omega, \mathcal{A})$, $\{X_n\}$ are independent with densities $\{f_n\}$ and $\{g_n\}$ under P and Q respectively. Suppose for each n, (6.27) holds. Then on $\mathcal{G} = \vee_{n=1}^{\infty} \mathcal{G}_n$, $Q \ll P$ if and only if $\prod_{n\geq 1} \int \sqrt{f_n(x)g_n(x)} \lambda(dx) > 0$. Moreover, this condition is equivalent to $\sum_{n\geq 1} \int \left(\sqrt{f_n(x)} - \sqrt{g_n(x)}\right)^2 \lambda(dx) < \infty.$* ◆

Proof Recall that on $\mathcal{G}_n$, $\frac{dQ}{dP} = L_n$. The candidate Radon–Nikodym derivative for $\frac{dQ}{dP}$ on $\mathcal{G}$ must be the almost sure limit L_∞ of L_n. But then, since Q is a probability measure on $\mathcal{G}$, we require $\mathrm{E}_P(L_\infty) = 1$.

By Kakutani's Theorem 6.7.1, $(L_n, n \geq 1)$ is uniformly integrable (and $\mathrm{E}(L_\infty) = 1$) if and only if

$$\prod_{n\geq 1} \mathrm{E}_P \left(\sqrt{\frac{g_n(X_n)}{f_n(X_n)}}\right) > 0. \tag{6.28}$$

Now note that

$$\mathrm{E}_P \left(\sqrt{\frac{g_n(X_n)}{f_n(X_n)}}\right) = \int_\mathbb{R} f_n(x) \sqrt{\frac{g_n(x)}{f_n(x)}} \lambda(dx)$$
$$= \int_\mathbb{R} \sqrt{f_n(x) g_n(x)} \lambda(dx).$$

Since $(\sqrt{f(x)} - \sqrt{g(x)})^2 = f(x) + g(x) - 2\sqrt{f(x)g(x)}$, it follows that

$$\prod_{n\geq 1} \int \sqrt{f_n(x)g_n(x)} \lambda(dx) > 0 \Leftrightarrow \sum_{n\geq 1} \int (\sqrt{f_n(x)} - \sqrt{g_n(x)})^2 \lambda(dx) < \infty.$$

□

Example 6.8.1 Let $\{X_n\}$ be a sequence of iid random variables on $(\Omega, \mathcal{A})$ under probability measures P and Q. Suppose that the induced measure of X_1 has density $f(\cdot)$ and $g(\cdot)$ under P and Q respectively where $g(x) = 0$ whenever $f(x) = 0$. Then by Theorem 6.8.2

$$Q \ll P \text{ on } \mathcal{G} = \sigma(X_1, X_2, \ldots) \text{ if and only if } \int (\sqrt{f(x)} - \sqrt{g(x)})^2 dx = 0,$$

that is, $f = g$ almost surely.

This has applications to statistical experiments. Let $X_1, \ldots, X_n$ be iid observations. Consider the problem of accepting/rejecting one of the following two hypotheses:

$$H_0 : X_1 \text{ has density } f(\cdot) \text{ against } H_1 : X_1 \text{ has density } g(\cdot).$$

We employ the following test: If $L_n = \prod_{i=1}^n \frac{g_i(X_i)}{f_i(X_i)} < 1$ then accept H_0, otherwise reject H_0. This test is "consistent"—if H_0 is true, then $L_n \to 0$ almost surely; while if H_1 is true, then $L_n \to \infty$ almost surely. ▲

6.9 Exercises

Exercise 6.9.1 Suppose that $X_1, X_2, \ldots$ are iid random variables and $S_n := \sum_{i=1}^n X_i, n \geq 1$.
(a) Show that $P\left(\limsup_{n\to\infty} S_n > 0\right) = 0$ or 1.
(b) Does (a) follow from Kolmogorov's 0-1 law?

Exercise 6.9.2 Given an example of a r.v. which is measurable with respect to the permutation invariant σ-field but is not measurable with respect to the tail σ-field.

Exercise 6.9.3 Suppose $\{X_n\}_{n\geq 1}$ is a sequence of exchangeable random variables with $E(X_1^2) < \infty$. Show that $\text{Cov}(X_i, X_j) \geq 0$.

Exercise 6.9.4 Let $\{X_i\}_{i\geq 1}$ be an exchangeable sequence of random variables with values in $\{0, 1\}$. Show that $\lim_{n\to\infty} E(X_{n+1}|X_1, \ldots, X_n)$ exists almost surely.

Exercise 6.9.5 Let U be a random variable on $(\Omega, \mathcal{A}, P)$ which has a uniform distribution on the interval $(0, 1)$.

(a) Show that on this probability space, there exist iid random variables $\{X_n\}$, where each X_n is $\sigma(U)$-measurable and has a Bernoulli$(1/2)$ distribution. *Hint*: Consider the binary expansion of U.

(b) Show that on this probability space, there exist iid random variables $\{U_i\}$, which are $\sigma(U)$-measurable, and are uniformly distributed on $(0, 1)$. *Hint*: Split $\{X_1, X_2, \ldots\} = \cup A_i$ where each A_i countably infinite.

(c) Let $\{F_i\}$ be a sequence of CDFs. Construct $\sigma(U)$-measurable independent $\{Z_i\}$ such that for each n, F_n is the CDF of Z_n.

Exercise 6.9.6 Let $(\Omega, \mathcal{A})$ be a measurable space and $\mathcal{A}_1, \mathcal{A}_2, \ldots$ be σ-fields with $\mathcal{A} = \bigvee_{n=1}^{\infty} \mathcal{A}_n$. There are probability measures P, Q on $(\Omega, \mathcal{A})$ such that $\mathcal{A}_1, \mathcal{A}_2, \ldots$ are independent under both P and Q. Further,

$$P(A) = 0 \iff Q(A) = 0 \text{ for all } A \in \bigcup_{n=1}^{\infty} \mathcal{A}_n.$$

(a) Give an example to show that $P \ll Q$ is not necessarily true.

(b) Prove that $P \ll Q$ if and only if $Q \ll P$. *Hint*: Use Kakutani's theorem and Exercise 5.5.3.

Exercise 6.9.7 Outline of an alternative proof of de Finetti's Theorem 6.6.2 from Aldous (1985) which is similar to the earlier proof.

Suppose $\{Z_i\}$ is an infinite sequence of exchangeable real r.vs. Define σ-fields

$$\mathcal{G}_n = \sigma(f_n(Z_1, \ldots, Z_n) : f_n \text{ is symmetric and measurable}),$$
$$\mathcal{H}_n = \sigma(\mathcal{G}_n, Z_{n+1}, Z_{n+2}, \ldots).$$

(a) Show that for every Y which is $\mathcal{H}_n$-measurable,

$$(Z_i, Y) \stackrel{d}{=} (Z_1, Y), \quad 1 \leq i \leq n.$$

(First note that it holds for all Y of the form $(f_n(Z_1, \ldots, Z_n), Z_{n+1}, \ldots))$, f_n symmetric.)

(b) Show that for any bounded measurable ϕ,

$$E(\phi(Z_1)| \mathcal{H}_n) = \frac{1}{n} \sum_{i=1}^{n} \phi(Z_i) \text{ almost surely.}$$

6.9 Exercises

(c) Show that

$$\frac{1}{n}\sum_{i=1}^{n}\phi(Z_i) \to E(\phi(Z_1)|\mathcal{H}) \quad \text{a.s., where } \mathcal{H} = \bigcap_{n=1}^{\infty}\mathcal{H}_n. \tag{6.29}$$

(d) Show the following extension of (c). For any bounded measurable $\phi(x_1, \ldots, x_k)$, for all $n > k$, almost surely,

$$E(\phi(Z_1, \ldots, Z_k)|\mathcal{H}_n) = \frac{1}{\prod_{i=0}^{k-1}(n-i)} \sum_{\{j_r\}\in D_{k,n}} \phi(Z_{j_1}, \ldots, Z_{j_k}),$$

where

$$D_{k,n} = \{(j_1, \ldots, j_k) : 1 \le j_r \le n, \{j_r\} \text{ distinct}\}.$$

(e) Show that $\#D_{k,n} = O(n^{k-1})$, and

$$\frac{1}{n^k}\sum_{j_1=1}^{n}\cdots\sum_{j_k=1}^{n}\phi(Z_{j_1}, \ldots, Z_{j_k}) \to E(\phi(Z_1, \ldots, Z_k)|\mathcal{H}) \quad \text{a.s.}$$

(f) Conside ring $\phi(x_1, \ldots, x_k)$ of the form $\prod_{i=1}^{k}\phi_i(x_i)$, show that

$$E\left(\prod_{r=1}^{k}\phi_r(Z_r)\Big|\mathcal{H}\right) = \prod_{r=1}^{k}E(\phi_r(Z_r)|\mathcal{H}).$$

(g) Conclude that $\{Z_i\}$ are conditionally independent given $\mathcal{H}$.

(h) Argue that $\{Z_i\}$ are also identically distributed given $\mathcal{H}$.

Exercise 6.9.8 Suppose X and Y be real-valued random variables on $(\Omega, \mathcal{A}, P)$, and $\mathcal{F}$ is a sub-σ-field. Show that conditional independence of X and Y given $\mathcal{F}$ is equivalent to each of the three statements below:

For all bounded measurable functions ϕ, ϕ_1, ϕ_2 and $A \in \mathcal{A}$,

(E1) $E(\phi_1(X)\phi_2(Y)|\mathcal{F}) = E(\phi_1(X)|\mathcal{F})E(\phi_2(Y)|\mathcal{F})$.

(E2) $P(X \in A|\mathcal{F}, Y) = P(X \in A|\mathcal{F})$.

(E3) $E(\phi(X)|\mathcal{F}, Y) = E(\phi(X)|\mathcal{F})$.

Exercise 6.9.9 (a) If X and X are conditionally independent given $\mathcal{F}$ then show that $X \in \mathcal{F}$ a.s.

(b) Suppose for each $j \ge 1$, conditioned on $\mathcal{F}$, X_j and $\sigma(X_i : i > j)$ are conditionally independent. Then show that $\{X_i\}$ are conditionally independent given $\mathcal{F}$.

(c) Suppose X and $\mathcal{F}$ are conditionally independent given $\mathcal{G}$, and suppose X and $\mathcal{G}$ are conditionally independent given $\mathcal{H} \subseteq \mathcal{G}$. Then show that X and $\mathcal{F}$ are conditionally independent given $\mathcal{H}$.

Exercise 6.9.10 (*Another proof of de Finetti's theorem from* Aldous (1985)) Suppose $\{Z_i\}$ is an infinite sequence of exchangeable random variables. Let

$$\mathcal{F}_n := \sigma(Z_n, Z_{n+1}, \ldots),$$

$$\mathcal{T} := \bigcap_{n=1}^{\infty} \mathcal{F}_n \text{ (the tail } \sigma\text{-field)}.$$

Let $\phi : \mathbb{R} \to \mathbb{R}$ be bounded and measurable.

(a) Show that

$$\lim_{n \to \infty} E(\phi(Z_1)|\mathcal{F}_n) = E(\phi(Z_1)|\mathcal{T}) \text{ a.s.},$$

$$E(\phi(Z_1)|\mathcal{F}_2) \stackrel{d}{=} E(\phi(Z_1)|\mathcal{T}),$$

and hence Z_1 and $\mathcal{F}_2$ are conditionally independent given $\mathcal{T}$.

(b) Show that for $m \geq 1$, Z_m and $\mathcal{F}_{m+1}$ are conditionally independent given $\mathcal{T}$.

(c) Show that $\{Z_i\}$ are conditionally independent given $\mathcal{T}$.

(d) Show that for every n, $E(\phi(Z_1)|\mathcal{F}_{n+1}) = E(\phi(Z_n)|\mathcal{F}_{n+1})$ a.s. for every bounded measurable ϕ.

(e) Show that $E(\phi(Z_1)|\mathcal{T}) = E(\phi(Z_n)|\mathcal{T})$ a.s. Thus conditioned on $\mathcal{T}$, $\{Z_i\}$ are identically distributed. Hence $\{Z_i\}$ are iid conditioned on $\mathcal{T}$.

References

D.J. Aldous, Exchangeability and related topics, in *École d'Été de Probabilités de Saint-Flour XIII - 1983*, ed. by P.L. Hennequin (Springer, Berlin Heidelberg, 1985), pp. 1–198

R. Berk, Limiting behavior of posterior distributions where the model is incorrect. Ann. Math. Stat. **37**, 51–58 (1966)

B. de Finetti, *Probability Induction and Statistics. The Art of Guessing* (John Wiley & Sons, London-New York-Sydney, 1972)

E. Hewitt, L.J. Savage, Symmetric measures on Cartesian products. Trans. Am. Math. Soc. **80**, 470–501 (1955)

W. Kirsch, An elementary proof of de Finetti's theorem. Stat. Probab. Lett. **151**, 84–88 (2019)

Chapter 7
Central Limit Theorem

Convergence in distribution results for sums of random variables is generally known as Central limit theorems (CLTs). They are central in probability theory and its applications. The literature is very rich in such results, obtained under different conditions on the variables. We first state two classical CLTs for independent random variables. Then we state and prove a CLT for martingales. Finally, we discuss easy applications of the martingale CLT to urn models, high-dimensional random matrices, and Markov chains.

7.1 Central Limit Theorem: Independent Summands

In Exercise 1.11.5, we saw that a suitably centered and scaled binomial random variable converges in distribution to a Gaussian distribution. On the other hand, a binomial random variable can be written as a sum of iid (Bernoulli) random variables. The following two historic results provide sweeping generalizations of the Binomial case. Since we shall soon state and prove a more general result, we omit their proofs. The first result is for iid summands, and the next result is an extension due to Lindeberg (1922), Lévy (1937).

Theorem 7.1.1 (Central limit theorem (CLT)) *Let $\{X_i\}$ be iid random variables with $E(X_1) = 0$ and $E(X_1^2) = 1$. Then as $n \to \infty$, $n^{-1/2}(X_1 + \cdots + X_n) \Rightarrow Z \sim N(0, 1)$.* ◆

Theorem 7.1.2 (Lévy–Lindeberg CLT) *Suppose that for each n, $\{Y_{ni}\}_{1 \le i \le k_n}$, $k_n \to \infty$, are independent zero mean random variables such that $\lim_{n \to \infty} \sum_{i=1}^{k_n} E(Y_{ni}^2) = \sigma^2 > 0$, and*

$$\lim_{n \to \infty} \sum_{i=1}^{k_n} E\left(Y_{ni}^2 \mathbf{1}_{\{|Y_{ni}| > \varepsilon\}}\right) = 0, \text{ for all } \varepsilon > 0. \tag{7.1}$$

Then as $n \to \infty$, $\sum_{i=1}^{k_n} Y_{ni} \Rightarrow Z \sim N(0, \sigma^2)$. ◆

Exercise 7.1.1 Show that Theorem 7.1.2 implies Theorem 7.1.1.

Exercise 7.1.2 Show that (7.1) holds if we assume that, for some $\delta > 0$,

$$\lim_{n \to \infty} \sum_{i=1}^{k_n} E\big(|Y_{ni}|^{2+\delta}\big) = 0.$$

7.2 Martingale Central Limit Theorem

Example 7.2.1 Let $\{\varepsilon_n\}_{n \geq 1}$ be iid taking values ± 1 with probability $1/2$ each. Suppose $W_1 = 0$, and for $n \geq 2$,

$$W_n = \begin{cases} 1 & \text{if } \varepsilon_1 = 1, \\ 2 & \text{if } \varepsilon_1 = -1. \end{cases}$$

Let $X_n = \sum_{i=1}^n W_i \varepsilon_i = \sum_{i=2}^n W_i \varepsilon_i$. It is easy to check that $\{X_n\}$ is a martingale. Let $S_n = \varepsilon_2 + \cdots + \varepsilon_n$ for $n \geq 2$. Note that $X_n = S_n$ or $2S_n$ depending on whether $\varepsilon_1 = +1$ or $\varepsilon_1 = -1$. Using this, and Theorem 7.1.1, it is easy to check that X_n converges in distribution. However, the limit X is not Gaussian but rather a "mixture" of two Gaussian distributions with means 0 and standard deviations 1 and 2. ▲

It is clear from Example 7.2.1 that the distributional limits of martingales need not be Gaussian. There are extremely general martingale CLT where the limits are quite complicated. See for example Hall and Heyde (1980). We focus only on a relatively simple situation where the limit is Gaussian. This result is very similar in spirit to the Lévy–Lindeberg CLT.

Theorem 7.2.1 (Martingale CLT) *Suppose that* $(S_{nk}, \mathcal{G}_{nk})_{k \geq 0}$ *with* $S_{n0} = 0$ *is a martingale for each* $n \in \mathbb{N}$. *Let* $Y_{nk} = S_{nk} - S_{n\,k-1}$, $k \geq 1$. *Assume that*

$$\sum_{k=1}^\infty E(Y_{nk}^2) < \infty, \tag{7.2}$$

and

$$\lim_{n \to \infty} \sum_{k=1}^\infty E\big(Y_{nk}^2 \mathbf{1}_{\{|Y_{nk}| \geq \varepsilon\}}\big) = 0, \text{ for all } \varepsilon > 0, \tag{7.3}$$

and for some constant $\sigma > 0$,

$$\sum_{k=1}^\infty E(Y_{nk}^2 | \mathcal{G}_{n\,k-1}) \xrightarrow{P} \sigma^2, \text{ as } n \to \infty. \tag{7.4}$$

7.2 Martingale Central Limit Theorem

Then,

$$\sum_{k=1}^{\infty} Y_{n,k} \Rightarrow Z \sim N(0, \sigma^2) \text{ as } n \to \infty.$$

♦

Proof of Theorem 7.2.1 First note that for any fixed n, (7.2) implies that $\sum_{j=1}^{k} Y_{nj} = S_{nk}$, as $k \to \infty$, converges to some S_n almost surely by an easy application of Theorem 5.3.1 with $p = 2$, see Exercise 5.5.6. That is, $S_n = \sum_{k=1}^{\infty} Y_{nk}$.

Let the conditional variances and their partial sums be denoted by

$$\sigma_{nk}^2 := \mathrm{E}(Y_{nk}^2 | \mathcal{G}_{n\,k-1}), \ n, k \in \mathbb{N},$$

$$\Sigma_{nk} := \sum_{j=1}^{k} \sigma_{nj}^2, \ n \in \mathbb{N}, \ 1 \le k \le \infty.$$

Note that condition (7.2) ensures that for every n,

$$\sum_{k=1}^{\infty} Y_{nk}^2 < \infty \text{ and } \sum_{k=1}^{\infty} \sigma_{nk}^2 < \infty \text{ almost surely}. \tag{7.5}$$

For the time being, we impose an additional restriction which shall be removed later. Assume that for some constant C,

$$\Sigma_{n\infty} = \sum_{k=1}^{\infty} \sigma_{nk}^2 \le C < \infty, \text{ for all } n \ge 1. \tag{7.6}$$

We first show that $S_n \Rightarrow Z$.

By Lévy continuity Theorem 1.11.1, it suffices to show that

$$\lim_{n \to \infty} \mathrm{E}(e^{\iota t S_n}) = e^{-t^2 \sigma^2/2}, \text{ for all } t \in \mathbb{R}, \tag{7.7}$$

where $\iota = \sqrt{-1}$.

Fix t and observe that

$$\left| \mathrm{E}(e^{\iota t S_n}) - e^{-t^2\sigma^2/2} \right|$$
$$= \left| \mathrm{E}\left(e^{\iota t S_n} - e^{\iota t S_n + (\Sigma_{n\infty} - \sigma^2)t^2/2}\right) + \mathrm{E}\left(e^{\iota t S_n + (\Sigma_{n\infty} - \sigma^2)t^2/2} - e^{-t^2\sigma^2/2}\right) \right|$$
$$\le \mathrm{E}\left(\left|e^{\iota t S_n}\left(1 - e^{(\Sigma_{n\infty} - \sigma^2)t^2/2}\right)\right|\right) + e^{-t^2\sigma^2/2}\left|\mathrm{E}\left(e^{\iota t S_n + \Sigma_{n\infty} t^2/2}\right) - 1\right|$$
$$\le \mathrm{E}\left(\left|1 - e^{(\Sigma_{n\infty} - \sigma^2)t^2/2}\right|\right) + \left|\mathrm{E}\left(e^{\iota t S_n + \Sigma_{n\infty} t^2/2}\right) - 1\right|.$$

By Conditions (7.4) and (7.6), $(\Sigma_{n\infty} - \sigma^2)$ remains uniformly bounded as $n \to \infty$, and converges to 0 in probability. Hence, by Theorem 1.5.1(iv), the first term on the right side above converges to 0. Thus (7.7) will follow if we can show that

$$\lim_{n \to \infty} \mathrm{E}\left(e^{\iota t S_n + \Sigma_{n\infty} t^2/2}\right) = 1. \tag{7.8}$$

Proceeding toward the above, fix n.

Again using (7.6), note that

$$\lim_{k \to \infty} \mathrm{E}\left(e^{\iota t S_{nk} + \Sigma_{nk} t^2/2}\right) = \mathrm{E}\left(e^{\iota t S_n + \Sigma_{n\infty} t^2/2}\right).$$

This implies

$$\mathrm{E}\left(e^{\iota t S_n + \Sigma_{n\infty} t^2/2} - 1\right)$$
$$= \sum_{k=1}^{\infty} \mathrm{E}\left(e^{\iota t S_{nk} + \Sigma_{nk} t^2/2} - e^{\iota t S_{n\,k-1} + \Sigma_{n\,k-1} t^2/2}\right)$$
$$= \sum_{k=1}^{\infty} \mathrm{E}\left(e^{\iota t S_{n\,k-1}} e^{\Sigma_{nk} t^2/2} \mathrm{E}\left(e^{\iota t Y_{nk}} - e^{-\sigma_{nk}^2 t^2/2} \big| \mathcal{G}_{n\,k-1}\right)\right).$$

Hence, again in view of (7.6),

$$\left|\mathrm{E}\left(e^{\iota t S_n + \Sigma_{n\infty} t^2/2} - 1\right)\right| \leq e^{Ct^2} \sum_{k=1}^{\infty} \mathrm{E}\left(\left|\mathrm{E}\left(e^{\iota t Y_{nk}} - e^{-\sigma_{nk}^2 t^2/2} \big| \mathcal{G}_{n\,k-1}\right)\right|\right). \tag{7.9}$$

Fix $k \in \mathbb{N}$. Taylor's expansion implies that

$$e^{\iota t Y_{nk}} = 1 + \iota t Y_{nk} - \frac{1}{2} t^2 Y_{nk}^2 + \theta,$$

where θ is a measurable function and

$$|\theta| \leq |tY_{nk}|^3 \wedge (tY_{nk})^2.$$

Fix $\varepsilon > 0$ and let

$$\mathbf{1}_{nk} := \mathbf{1}_{\{|Y_{nk}| \geq \varepsilon\}}.$$

Thus, there exists a constant K_t, depending only on t, such that

$$|\theta| \leq K_t \left(|Y_{nk}|^3 \wedge Y_{nk}^2\right)$$
$$\leq K_t \left(Y_{nk}^2 \mathbf{1}_{nk} + |Y_{nk}|^3(1 - \mathbf{1}_{nk})\right)$$
$$\leq K_t \left(Y_{nk}^2 \mathbf{1}_{nk} + \varepsilon Y_{nk}^2\right). \tag{7.10}$$

7.2 Martingale Central Limit Theorem

Taylor's expansion also implies that

$$e^{-\sigma_{nk}^2 t^2/2} = 1 - \frac{1}{2}t^2\sigma_{nk}^2 + \theta',$$

where θ' is measurable, and for some constant $\bar{K}_t \geq K_t$,

$$|\theta'| \leq \bar{K}_t \sigma_{nk}^4. \tag{7.11}$$

To tackle the individual terms in (7.9), recall that

$$E(Y_{nk}|\mathcal{G}_{n\,k-1}) = 0 \text{ and } E(Y_{nk}^2|\mathcal{G}_{n\,k-1}) = \sigma_{nk}^2.$$

Hence, using (7.10) and (7.11),

$$\begin{aligned}
&\left|E\left(e^{itY_{nk}} - e^{-\sigma_{nk}^2 t^2/2}\big|\mathcal{G}_{n\,k-1}\right)\right| \\
&= \Big|E\big(e^{itY_{nk}} - 1 - itY_{nk} + \frac{1}{2}t^2 Y_{nk}^2\big|\mathcal{G}_{n\,k-1}\big) \\
&\quad - E\big(e^{-\sigma_{nk}^2 t^2/2} - 1 + \frac{1}{2}t^2\sigma_{nk}^2\big|\mathcal{G}_{n\,k-1}\big)\Big| \\
&\leq E\big(|\theta| + |\theta'|\big|\mathcal{G}_{n\,k-1}\big) \leq \bar{K}_t\big[E(Y_{nk}^2 \mathbf{1}_{nk}|\mathcal{G}_{n\,k-1}) + \varepsilon\sigma_{nk}^2 + \sigma_{nk}^4\big].
\end{aligned}$$

Note that

$$\begin{aligned}
\sigma_{nk}^2 = E(Y_{nk}^2|\mathcal{G}_{n\,k-1}) &\leq \varepsilon^2 + E(Y_{nk}^2 \mathbf{1}_{nk}|\mathcal{G}_{n\,k-1}) \\
&\leq \varepsilon^2 + \sum_{j=1}^{\infty} E(Y_{nj}^2 \mathbf{1}_{nj}|\mathcal{G}_{n\,j-1}),
\end{aligned}$$

and hence

$$E\big(\sup_{k\geq 1} \sigma_{nk}^2\big) \leq \varepsilon^2 + \sum_{k=1}^{\infty} E(Y_{nk}^2 \mathbf{1}_{nk}). \tag{7.12}$$

Now going back to (7.9),

$$\begin{aligned}
&\left|E\left(e^{itS_n + \Sigma_{n\infty}t^2/2} - 1\right)\right| \\
&\leq \bar{K}_t e^{Ct^2} \sum_{k=1}^{\infty} E\big(Y_{nk}^2 \mathbf{1}_{nk} + \varepsilon\sigma_{nk}^2 + \sigma_{nk}^4\big) \\
&\leq \bar{K}_t e^{Ct^2} E\Big(\sum_{k=1}^{\infty} Y_{nk}^2 \mathbf{1}_{nk} + \big(\varepsilon + \sup_{k\geq 1}\sigma_{nk}^2\big)\Sigma_{n\infty}\Big)
\end{aligned}$$

$$\leq \bar{K}_t e^{Ct^2} \Big[\sum_{k=1}^{\infty} E(Y_{nk}^2 \mathbf{1}_{nk}) + C\big(\varepsilon + E(\sup_{k\geq 1} \sigma_{nk}^2)\big) \Big]$$

$$\leq \bar{K}_t e^{Ct^2} \Big[(1+C) \sum_{k=1}^{\infty} E(Y_{nk}^2 \mathbf{1}_{nk}) + C(\varepsilon + \varepsilon^2) \Big], \tag{7.13}$$

where the last line follows from (7.12). Letting $n \to \infty$ in (7.13) and using (7.3), we have

$$\limsup_{n\to\infty} \big| E\big(e^{itS_n + \Sigma_{n\infty} t^2/2} - 1\big) \big| \leq \bar{K}_t e^{Ct^2} (\varepsilon + \varepsilon^2) C.$$

Since ε was arbitrary, left side of (7.9) goes to zero as $n \to \infty$, that is, (7.8) holds. This, in turn, implies (7.7). Thus $S_n \Rightarrow Z$ under the additional restriction (7.6).

If (7.6) does not hold, fix $C > \sigma^2$, and let

$$A_{nk} := \Big\{ \omega : \sum_{j=1}^{k} \sigma_{nj}^2(\omega) \leq C \Big\}, \quad Z_{nk} := Y_{nk} \mathbf{1}_{A_{nk}}, \quad 1 \leq n < \infty, 1 \leq k \leq \infty.$$

Since $\mathbf{1}_{A_{nk}} \in \mathcal{G}_{n\,k-1}$, it follows that

$$E(Z_{nk} | \mathcal{G}_{n\,k-1}) = 0, \quad \text{for } k, n \in \mathbb{N}.$$

Let

$$\theta_{nk}^2 = E(Z_{nk}^2 | \mathcal{G}_{n\,k-1}) = \sigma_{nk}^2 \mathbf{1}_{A_{nk}}.$$

Observe that for n fixed, $A_{nk} \downarrow A_{n\infty}$ as $k \to \infty$, and hence

$$\Omega = A_{n\infty} \cup \Big(\bigcup_{j=1}^{\infty} (A_{n\,j-1} \setminus A_{nj}) \Big),$$

where $A_{n0} = \Omega$. So, for any $\omega \in \Omega$, either $\omega \in A_{n\infty}$ or $\omega \in A_{n\,j-1} \setminus A_{nj}$ for some $j \in \mathbb{N}$.

Further,

$$\sum_{j=1}^{\infty} \theta_{nj}^2(\omega) = \begin{cases} \sum_{j=1}^{k} \sigma_{nj}^2(\omega) & \text{if } \omega \in A_{nk} \setminus A_{n\,k+1} \text{ for some } k \geq 0, \\ \sum_{j=1}^{\infty} \sigma_{nj}^2(\omega) & \text{if } \omega \in A_{n\infty}. \end{cases}$$

In any case, $\sum_{j=1}^{\infty} \theta_{nj}^2(\omega) \leq C$ holds true. Besides,

$$\mathbf{1}_{A_{n\infty}} \sum_{j=1}^{\infty} \theta_{nj}^2 = \mathbf{1}_{A_{n\infty}} \sum_{j=1}^{\infty} \sigma_{nj}^2.$$

7.2 Martingale Central Limit Theorem

On the other hand, $C > \sigma^2$ and (7.4) together imply that $\mathbf{1}_{A_{n\infty}} \xrightarrow{P} 1$ as $n \to \infty$. Therefore,

$$\lim_{n\to\infty} \sum_{j=1}^{\infty} \theta_{nj}^2 \xrightarrow{P} \sigma^2.$$

By the special case that has already been proved, $\sum_{k=1}^{\infty} Z_{nk} \Rightarrow Z$. Now note that,

$$S_n = S_n \mathbf{1}_{A_{n\infty}} + S_n \mathbf{1}_{A_{n\infty}^c} = \mathbf{1}_{A_{n\infty}} \sum_{k=1}^{\infty} Z_{nk} + S_n \mathbf{1}_{A_{n\infty}^c}.$$

It easily follows, for example, by using Exercise 1.10.11, that $S_n \Rightarrow Z$ and the proof is complete. □

Check that Theorem 7.1.2 follows from Theorem 7.2.1. Statement and proof of Theorem 7.2.1 remain essentially unchanged if, for some n, $(S_{nk}, \mathcal{G}_{nk})_{k\geq 0}$ is replaced by $(S_{nk}, \mathcal{G}_{nk})_{0\leq k\leq k_n}$, for some $k_n \geq 1$, and $\sum_{k=1}^{\infty}$ is replaced by $\sum_{k=1}^{k_n}$ everywhere. The following consequence of Theorem 7.2.1 is often useful as its conditions are easy to check.

Theorem 7.2.2 *Suppose $(X_n, \mathcal{G}_n)_{n\geq 0}$ is a martingale with $X_0 = 0$. Let*

$$A_n = \sum_{k=1}^{n} \mathrm{E}\left((X_k - X_{k-1})^2 | \mathcal{G}_{k-1}\right), n \geq 1,$$

be the quadratic variation of (X_n) as in Exercise 4.7.5. If for some $\theta, \sigma_n > 0$, $\sigma_n^{-2} A_n \xrightarrow{P} \theta^2$ as $n \to \infty$, and for some $\delta > 0$,

$$\lim_{n\to\infty} \sigma_n^{-(2+\delta)} \sum_{k=1}^{n} \mathrm{E}\left(|X_k - X_{k-1}|^{2+\delta}\right) = 0,$$

then as $n \to \infty$, $\sigma_n^{-1} X_n \Rightarrow Z \sim N(0, \theta^2)$. ◆

Proof Define $S_{nk} := \sigma_n^{-1} X_k$ and $\mathcal{G}_{nk} := \mathcal{G}_k$, $0 \leq k \leq n$. Then we can directly use Exercise 7.1.2 and Theorem 7.2.1. □

A few applications of the martingale central limit theorems to urn models, random matrices, and Markov chains are presented in the next three sections. As illustration is the purpose, we leave out the most general results known, and present only the simplest cases.

7.3 Urn Model

Consider an urn which has W_0 white balls and B_0 black balls. A ball is drawn at random from the urn with replacement, and α balls of the color drawn and β balls of the opposite color are added to the urn, where $\alpha, \beta \geq 0$ are integers. This is repeated ad infinitum.

When $\beta = 0$, this is Pólya's urn model, considered in Example 5.3.1. The general case was studied in Friedman (1949).

Let W_n and B_n be the number of white balls and black balls, respectively, in the urn after n draws. In the case $\beta \geq 1$ and

$$\rho = \frac{\alpha - \beta}{\alpha + \beta} < \frac{1}{2},$$

Freedman (1965) showed that as $n \to \infty$,

$$n^{-1/2}(W_n - B_n) \Rightarrow Z \sim N\left(0, (\alpha - \beta)^2/(1 - 2\rho)\right). \tag{7.14}$$

We now show how the martingale CLT implies (7.14). For the sake of simplicity, we consider the case $\alpha = 0$, $\beta = 1$, $W_0 = 1$, $B_0 = 1$. Theorem 7.3.1 is clearly consistent with (7.14) for this pair of (α, β).

Theorem 7.3.1 *As* $n \to \infty$, $n^{-1/2}(W_n - B_n) \Rightarrow Z \sim N\left(0, \frac{1}{3}\right).$ ◆

The following observation, which is essentially a restatement of Exercise 4.7.5, is needed for the proof.

Exercise 7.3.1 Let $(X_n, \mathcal{G}_n)_{n \geq 0}$ be a martingale, with $\mathrm{E}(X_{n+1}^2) < \infty$ for some $n \geq 0$. Then,

$$\mathrm{E}\left((X_{n+1} - X_n)^2\right) < \infty, \text{ and } \mathrm{E}\left((X_{n+1} - X_n)^2 | \mathcal{G}_n\right) = \mathrm{E}(X_{n+1}^2 | \mathcal{G}_n) - X_n^2.$$

Proof of Theorem 7.3.1 The assumptions $\alpha = 0$ and $\beta = 1$ mean, for example, if the first, second, and third draws yield, respectively, a white, a black, and a white ball, then $W_1 = 1$ and $W_2 = W_3 = 2$. Besides, $B_n = n + 2 - W_n$ for all n.

Let $\mathcal{G}_n = \sigma(W_0, W_1, \ldots, W_n)$, for $n \geq 0$. After n draws, the urn has $(n + 2)$ balls, of which W_n are white. Therefore,

$$P(W_{n+1} = W_n | \mathcal{G}_n) = (n + 2)^{-1} W_n, \quad n \geq 0. \tag{7.15}$$

Since $W_{n+1} - W_n$ is either 0 or 1, the above implies

$$\mathrm{E}(W_{n+1} | \mathcal{G}_n) = 1 + \frac{n + 1}{n + 2} W_n, \quad n \geq 0. \tag{7.16}$$

7.3 Urn Model

Letting
$$M_n = (n+1)W_n - \frac{1}{2}(n+1)(n+2), \ n \geq 0, \tag{7.17}$$

Equation (7.16) is equivalent to saying that $(M_n, \mathcal{G}_n)_{n \geq 0}$ is a martingale. For $n \geq 0$, write
$$M_{n+1} - M_n = (n+2)(W_{n+1} - W_n) + W_n - (n+2). \tag{7.18}$$

Given $\mathcal{G}_n$, the right side takes values $W_n - n - 2$ and W_n with probabilities $(n+2)^{-1}W_n$ and $1 - (n+2)^{-1}W_n$, respectively, by (7.15). Thus,

$$\begin{aligned}
E\big((M_{n+1} - M_n)^2 | \mathcal{G}_n\big) &= (W_n - n - 2)^2 \frac{W_n}{n+2} + W_n^2 \Big(1 - \frac{W_n}{n+2}\Big) \\
&= W_n(n + 2 - W_n).
\end{aligned} \tag{7.19}$$

Now, we claim that
$$M_n/(n+1)(n+2) \to 0 \text{ in } L^2. \tag{7.20}$$

To see this, using $M_0 = 0$,

$$\begin{aligned}
E(M_n^2) &= E\Big(\sum_{i=0}^{n-1} E(M_{i+1}^2 - M_i^2 | \mathcal{G}_i)\Big) \\
&= E\Big(\sum_{i=0}^{n-1} E\big((M_{i+1} - M_i)^2 | \mathcal{G}_i\big)\Big) \text{ (by Exercise 7.3.1)} \\
&= E\Big(\sum_{i=0}^{n-1} W_i(i+2-W_i)\Big) \text{ (by (7.19))} \\
&\leq \sum_{i=0}^{n-1} (i+2)^2 \text{ (since } 0 \leq W_i \leq i+2, \ i \geq 0\text{),}
\end{aligned}$$

from which, (7.20) immediately follows.

In view of (7.17), (7.20) implies that

$$\lim_{n \to \infty} \frac{W_n}{n+2} = \frac{1}{2} \text{ in } L^2. \tag{7.21}$$

Use (7.19) again to get

$$E\left|\sum_{i=0}^{n-1}\left[E((M_{i+1}-M_i)^2|\mathcal{G}_i) - \frac{(i+2)^2}{4}\right]\right|$$

$$= E\left|\sum_{i=0}^{n-1}(i+2)(W_i - \frac{i+2}{2}) + (\frac{(i+2)^2}{4} - W_i^2)\right|$$

$$\leq \sum_{i=0}^{n-1}(i+2)^2\left[E\left(\left|\frac{W_i}{i+2} - \frac{1}{2}\right| + \left|\frac{W_i^2}{(i+2)^2} - \frac{1}{4}\right|\right)\right].$$

Exercise 1.6.1(c) and Theorem 3.4.2 applied to (7.21) imply

$$\lim_{i\to\infty}\left(E\left|\frac{W_i}{i+2} - \frac{1}{2}\right| + E\left|\frac{W_i^2}{(i+2)^2} - \frac{1}{4}\right|\right) = 0.$$

Therefore,

$$\lim_{n\to\infty} n^{-3} E\left|\sum_{i=0}^{n-1} E((M_{i+1}-M_i)^2|\mathcal{G}_i) - \frac{1}{4}\sum_{i=0}^{n-1}(i+2)^2\right| = 0.$$

This in conjunction with $\lim_{n\to\infty} n^{-3}\sum_{i=0}^{n-1}(i+2)^2 = \frac{1}{3}$ establishes that

$$\lim_{n\to\infty} n^{-3}\sum_{i=0}^{n-1} E((M_{i+1}-M_i)^2|\mathcal{G}_i) = \frac{1}{12} \quad \text{in } L^1. \tag{7.22}$$

Recall (7.18), and that $W_{n+1} - W_n$ is either 0 or 1, and $|W_n| \leq n+2$ to conclude

$$|M_{n+1} - M_n| \leq 3(n+2), \; n \geq 0.$$

Therefore,

$$\sum_{i=0}^{n}(n^{-3/2}|M_{i+1}-M_i|)^3 \leq 27n^{-9/2}\sum_{i=0}^{n}(i+2)^3 \to 0, \text{ as } n \to \infty.$$

Thus, $\lim_{n\to\infty} n^{-9/2}\sum_{i=0}^{n} E(|M_{i+1}-M_i|^3) = 0$.

This and (7.22) ensure that the hypotheses of Theorem 7.2.2 hold. Hence, we can conclude that $n^{-3/2}M_n \Rightarrow Y \sim N(0, \frac{1}{12})$.

This is the same as saying that $n^{-1/2}(W_n - \frac{1}{2}(n+2)) \Rightarrow Y$. Recall that $W_n + B_n = n+2$, that is, $W_n - B_n = 2W_n - (n+2)$. Observing that $2Y \stackrel{d}{=} Z$, the proof is complete. □

7.4 Random Matrix

Random matrix theory is a favorite topic of the authors, who gave in to the temptation of including one example from this area. Let $\{X_{i,j} : 1 \leq i \leq j\}$ be a collection of iid random variables with mean zero, variance one, and finite fourth moment. For $N \geq 1$, let W_N be an $N \times N$ random symmetric matrix whose (i, j)-th entry is defined by

$$W_N(i, j) := X_{i \wedge j, i \vee j}, \ 1 \leq i, j \leq N.$$

In random matrix theory, W_N is called a *Wigner matrix*.

Let Tr denote the trace of a matrix. It is well known that under suitable conditions on the entries, for any integer $h \geq 1$, $\text{Tr}(W_N^h)$, after appropriate centering and normalization, converges to a Gaussian distribution.

While the proof of this result is beyond the scope of this book, in this section, we show how the martingale CLT can be used to prove the result for $h = 3$.

Theorem 7.4.1 *As $N \to \infty$, $N^{-3/2} \text{Tr}(W_N^3) \Rightarrow Z \sim N(0, 15)$.* ◆

Proof Write

$$\text{Tr}(W_N^3)$$
$$= \sum_{1 \leq i,j,k \leq N} W_N(i, j) W_N(j, k) W_N(k, i)$$
$$= \sum_{i=1}^{N} W_N(i, i)^3 + 3 \sum_{1 \leq i,j,k \leq N, i \neq j = k} W_N(i, j) W_N(j, k) W_N(k, i)$$
$$\quad + \sum_{1 \leq i,j,k \leq N, i,j,k \text{ distinct}} W_N(i, j) W_N(j, k) W_N(k, i)$$

$$= \sum_{i=1}^{N} W_N(i, i)^3 + 3 \sum_{1 \leq i < j \leq N} W_N(i, j)^2 W_N(j, j)$$
$$\quad + 3 \sum_{1 \leq j < i \leq N} W_N(i, j)^2 W_N(j, j)$$
$$\quad + 6 \sum_{1 \leq i < j < k \leq N} W_N(i, j) W_N(j, k) W_N(k, i)$$
$$= \sum_{i=1}^{N} X_{i,i}^3 + 3 \sum_{1 \leq i < j \leq N} X_{i,j}^2 (X_{i,i} + X_{j,j}) + 6 \sum_{1 \leq i < j < k \leq N} X_{i,j} X_{j,k} X_{i,k}.$$

Note the two relations,

$$\sum_{1\leq i<j\leq N} X_{i,j}^2 X_{i,i} = \sum_{1\leq i<j\leq N} (X_{i,j}^2 - 1)X_{i,i} + \sum_{i=1}^{N}(N-i)X_{i,i},$$

$$\sum_{1\leq i<j\leq N} X_{i,j}^2 X_{j,j} = \sum_{1\leq i<j\leq N} (X_{i,j}^2 - 1)X_{j,j} + \sum_{j=1}^{N}(j-1)X_{j,j}.$$

Thus,
$$\mathrm{Tr}\left(W_N^3\right) = 6M_N + 3U_N + 3V_N + Y_N, \quad N = 1, 2, \ldots, \qquad (7.23)$$

where

$$M_N = \sum_{1\leq i<j<k\leq N} X_{i,j} X_{j,k} X_{i,k}, \qquad (7.24)$$

$$U_N = (N-1)\sum_{i=1}^{N} X_{i,i}, \qquad (7.25)$$

$$V_N = \sum_{1\leq i<j\leq N} \left(X_{i,j}^2 - 1\right)\left(X_{i,i} + X_{j,j}\right), \qquad (7.26)$$

$$Y_N = \sum_{i=1}^{N} X_{i,i}^3. \qquad (7.27)$$

Consider the filtration

$$\mathcal{G}_0 = \{\emptyset, \Omega\}, \text{ and } \mathcal{G}_N = \sigma\{X_{i,j} : 1 \leq i \leq j \leq N\}, N \geq 1.$$

Then $(M_N, \mathcal{G}_N)$ is an adapted sequence.

Observe that,

$$M_{N+1} - M_N = \sum_{1\leq i<j\leq N} X_{i,j} X_{j,N+1} X_{i,N+1}. \qquad (7.28)$$

Since X_{ij}'s have mean zero, it easily follows that for $N \geq 1$, $\mathrm{E}\left(M_{N+1} - M_N | \mathcal{G}_N\right) = 0$. That is, $(M_N, \mathcal{G}_N)_{N\geq 0}$ is a martingale.

Observe that for $1 \leq i < j \leq N$ and $1 \leq i' < j' \leq N$, unless $i = i'$ and $j = j'$,

$$\mathrm{E}\left(X_{i,j} X_{j,N+1} X_{i,N+1} X_{i',j'} X_{j',N+1} X_{i',N+1} | \mathcal{G}_N\right) = 0.$$

Hence from (7.28) it follows that

$$\mathrm{E}\left((M_{N+1} - M_N)^2 | \mathcal{G}_N\right) = \sum_{1\leq i<j\leq N} X_{i,j}^2, \quad N \geq 1.$$

Therefore

7.4 Random Matrix

$$\sum_{k=1}^{N} \mathrm{E}\big((M_{k+1}-M_k)^2|\mathcal{G}_k\big) = \sum_{k=1}^{N} \sum_{1\le i<j\le k} X_{i,j}^2$$
$$= \sum_{1\le i<j\le N}(N-j+1)X_{i,j}^2. \qquad (7.29)$$

Thus,

$$N^{-3}\mathrm{E}\bigg(\sum_{k=1}^{N}\mathrm{E}\big((M_{k+1}-M_k)^2|\mathcal{G}_k\big)\bigg) = N^{-3}\sum_{j=0}^{N-1} j(N-j)$$
$$\to \int_0^1 x(1-x)\,dx = \frac{1}{6}. \qquad (7.30)$$

As $\{X_{ij} : 1 \le i \le j\}$ is an iid collection, (7.29) implies

$$\mathrm{Var}\bigg(\sum_{k=1}^{N}\mathrm{E}\big((M_{k+1}-M_k)^2|\mathcal{G}_k\big)\bigg) = \mathrm{Var}(X_{1,1}^2)\sum_{1\le i<j\le N}(N-j+1)^2$$
$$\le N^4\,\mathrm{Var}(X_{1,1}^2).$$

The assumption of finite fourth moment implies that $\mathrm{Var}(X_{1,1}^2) < \infty$. Thus,

$$\lim_{N\to\infty}\mathrm{Var}\bigg(N^{-3}\sum_{k=1}^{N}\mathrm{E}\big((M_{k+1}-M_k)^2|\mathcal{G}_k\big)\bigg) = 0,$$

which in conjunction with (7.30) shows

$$N^{-3}\sum_{k=1}^{N}\mathrm{E}\big((M_{k+1}-M_k)^2|\mathcal{G}_k\big) \xrightarrow{P} \frac{1}{6} \text{ as }, N\to\infty. \qquad (7.31)$$

Use (7.28) once again to get

$$\mathrm{E}\big((M_{N+1}-M_N)^4\big) = \sum_{\substack{1\le i_1<j_1\le N,\dots\\ \dots,1\le i_4<j_4\le N}}\mathrm{E}\bigg(\prod_{k=1}^{4}X_{i_k,j_k}X_{j_k,N+1}X_{i_k,N+1}\bigg)$$
$$= \sum_{\substack{1\le i_1<j_1\le N,\dots\\ \dots,1\le i_4<j_4\le N}}\mathrm{E}\bigg(\prod_{k=1}^{4}X_{i_k,j_k}\bigg)\mathrm{E}\bigg(\prod_{k=1}^{4}X_{j_k,N+1}X_{i_k,N+1}\bigg).$$

Consider the set

$$S = \left\{(i_1, j_1, \ldots, i_4, j_4) : i_k < j_k \leq N, k \leq 4 \text{ and } \mathrm{E}\Big(\prod_{k=1}^{4} X_{i_k, j_k}\Big) \neq 0\right\}. \quad (7.32)$$

Recall that X_{ij}'s have zero mean. This implies that $(i_1, j_1, \ldots, i_4, j_4) \in S$ only if either $i_1 = \cdots = i_4$, $j_1 = \cdots = j_4$, or there exists a permutation (k, l, m, n) of $(1, 2, 3, 4)$ such that $(i_k, j_k) = (i_l, j_l) \neq (i_m, j_m) = (i_n, j_n)$. This means there are $\binom{N}{2} + 3\binom{N}{2}\left(\binom{N}{2} - 1\right)$ elements in S. Further, for $(i_1, j_1, \ldots, i_4, j_4) \in S$, $\mathrm{E}(\prod_{k=1}^{4} X_{i_k, j_k})$ equals either $\mathrm{E}(X_{1,1}^4)$ or $\left(\mathrm{E}(X_{1,1}^2)\right)^2$, the latter being smaller than the former by Lyapunov's inequality (see Theorem 1.6.1). For the other factor, if $i_1 = \cdots = i_4$ and $j_1 = \cdots = j_4$, then

$$\mathrm{E}\Big(\prod_{k=1}^{4} X_{j_k, N+1} X_{i_k, N+1}\Big) = \left(\mathrm{E}(X_{1,1}^4)\right)^2. \quad (7.33)$$

On the other hand, if $(i_k, j_k) = (i_l, j_l) \neq (i_m, j_m) = (i_n, j_n)$ for some permutation (k, l, m, n) of $(1, 2, 3, 4)$, then

$$\mathrm{E}\Big(\prod_{k=1}^{4} X_{j_k, N+1} X_{i_k, N+1}\Big) = \begin{cases} \left(\mathrm{E}(X_{1,1}^2)\right)^4 & \text{if } \{i_j, i_l\} \cap \{i_m, j_m\} = \emptyset, \\ \left(\mathrm{E}(X_{1,1}^2)\right)^2 \mathrm{E}\left(X_{1,1}^4\right) & \text{otherwise}. \end{cases}$$

Lyapunov's inequality again shows the right side in either of the above cases is upper bounded by the right side of (7.33). Therefore

$$\mathrm{E}\left((M_{N+1} - M_N)^4\right) \leq 2N^4 \left(\mathrm{E}(X_{1,1}^4)\right)^3.$$

Hence

$$\lim_{N \to \infty} N^{-6} \sum_{k=1}^{N} \mathrm{E}\left((M_{k+1} - M_k)^4\right) = 0.$$

This together with (7.31) ensure that the hypotheses of Theorem 7.2.2 hold. Thus,

$$N^{-3/2} M_N \Rightarrow Z_1 \sim N\left(0, \frac{1}{6}\right). \quad (7.34)$$

Theorem 7.1.1 immediately implies that for U_N defined in (7.23), we have $N^{-3/2} U_N \Rightarrow Z_2 \sim N(0, 1)$. This, in view of independence of U_N and M_N, in conjunction with (7.34) proves that

$$\left(N^{-3/2} M_N, N^{-3/2} U_N\right) \Rightarrow (Z_1, Z_2), \quad (7.35)$$

where $Z_1 \sim N(0, \frac{1}{6})$, and $Z_2 \sim N(0, 1)$, and they are independent.

If V_N is as in (7.23), then $E(V_N) = 0$. For $1 \le i < j \le N$ and $1 \le i' < j' \le N$,

$$\text{Cov}\big((X_{i,j}^2 - 1)(X_{i,i} + X_{j,j}), (X_{i',j'}^2 - 1)(X_{i',i'} + X_{j',j'})\big) \text{ if } (i,j) \ne (i',j').$$

Thus $\text{Var}(V_N) = N(N-1)\text{Var}(X_{1,1}^2)$. Hence

$$N^{-3/2} V_N \xrightarrow{P} 0. \tag{7.36}$$

Finally, the SLLN Theorem 6.2.1 implies that $N^{-1} Y_N \to E(X_{1,1}^3)$ almost surely. Consequently, $N^{-3/2} Y_N \to 0$ almost surely. Combine this with (7.35) and (7.36) and use (7.23) to conclude that

$$N^{-3/2} \text{Tr}\left(W_N^3\right) \Rightarrow 6Z_1 + 3Z_2 \sim N(0, 15),$$

and this completes the proof. $\square$

7.5 Markov Chains

We shall now use the martingale CLT to obtain a CLT for Markov chains. We refer the reader to Norris (1998) for the definitions and elementary results related to Markov chains.

Let $(X_n : n \ge 1)$ be an *irreducible positive recurrent* Markov chain on a countable state space $\mathcal{S}$ with transition probabilities $(p_{ij} : i, j \in \mathcal{S})$ and initial probability distribution $(\mu\{i\} : i \in \mathcal{S})$. Let $k = \#\mathcal{S}$. Throughout this section, $\mathcal{S}$ is either finite ($k < \infty$) or is countably infinite ($k = \infty$). For notational simplicity, without loss of generality we take $\mathcal{S} = \{1, 2, \ldots, k\}$ if $k < \infty$, and $\mathcal{S} = \mathbb{N}$ if $k = \infty$.

By Theorem 1.7.7 of Norris (1998), (X_n) has an invariant distribution $(\pi_i : i \in \mathcal{S})$ with $\pi_i > 0$ for all i. That is, there exists $(\pi_i : i \in \mathcal{S})$ such that $\sum_{i=1}^k \pi_i = 1$ and

$$\sum_{i=1}^k \pi_i p_{ij} = \pi_j \text{ for all } j \in \mathcal{S}. \tag{7.37}$$

We now state the CLT for Markov chains.

Theorem 7.5.1 (CLT for Markov chains) *If $f : \mathcal{S} \to \mathbb{R}$ is such that*

$$\sum_{j=1}^k p_{ij} f(j) = 0 \text{ for all } i \in \mathcal{S}, \tag{7.38}$$

and

$$\sigma^2 = \sum_{j=1}^{k} \pi_j f(j)^2 < \infty, \qquad (7.39)$$

then as $n \to \infty$,

$$\frac{1}{\sqrt{n}} \sum_{i=1}^{n} f(X_i) \Rightarrow Z \sim N(0, \sigma^2).$$

◆

Proof We need to show that

$$\lim_{n \to \infty} P\left(n^{-1/2} \sum_{i=1}^{n} f(X_i) \le x\right) = P(Z \le x), \text{ for all } x \in \mathbb{R}. \qquad (7.40)$$

Suppose we are able to show that for all $a \in \mathcal{S}$,

$$\lim_{n \to \infty} P\left(n^{-1/2} \sum_{i=1}^{n} f(X_i) \le x \Big| X_1 = a\right) = P(Z \le x), \text{ for all } x \in \mathbb{R}. \qquad (7.41)$$

Then (7.40) would follow by an application of the DCT and the tower property. Therefore, we assume from now on that, $X_1 = a$, for a fixed $a \in \mathcal{S}$. That is, the initial distribution is the probability measure on $\mathcal{S}$ that puts mass 1 on a.

Let d be the period of $(X_n : n \ge 1)$. Use Theorem 1.8.4 of Norris (1998) to decompose $\mathcal{S} = C_1 \cup \cdots \cup C_d$ where $C_1, \ldots, C_d$ are disjoint, $a \in C_1$ and

$$X_{nd+r} \in C_r \text{ for all } r = 1, \ldots, d, \ n = 0, 1, 2, \ldots.$$

This result also implies that for all $1 \le r \le d, i, j \in C_r$, there exists $N_{ij} \in \mathbb{N}$ such that

$$p_{ij}^{(nd)} > 0, \ n \ge N_{ij}, \qquad (7.42)$$

where

$$p_{uv}^{(m)} = P(X_{m+1} = v | X_1 = u), \text{ for } m \ge 1 \text{ and } u, v \in \mathcal{S}.$$

Moreover, for every $1 \le r \le d$, $(X_{nd+r} : n \ge 0)$ is an irreducible Markov chain on C_r. Therefore, by Theorem 1.8.5 of Norris (1998), $(d\pi_i : i \in C_r)$ is an invariant distribution for $(X_{nd+r} : n \ge 0)$.

Let $(Y_n : n \ge 1)$ be a Markov chain, independent of $(X_n : n \ge 1)$, on $\mathcal{S}$ and transition probabilities $(p_{ij} : i, j \in \mathcal{S})$ and initial distribution,

$$P(Y_1 = i) = \begin{cases} d\pi_i & \text{if } i \in C_1, \\ 0 & \text{if } i \notin C_1. \end{cases}$$

7.5 Markov Chains

All this implies $Y_1 \stackrel{d}{=} Y_{d+1} \stackrel{d}{=} Y_{2d+1} \stackrel{d}{=} \cdots$. Inductively it follows that

$$Y_{nd+r} \stackrel{d}{=} Y_r, \ r = 1, \ldots, d, \ n = 1, 2, \ldots. \tag{7.43}$$

For a fixed $r = 1, \ldots, d$, $(Y_{nd+r} : n \geq 0)$ is an irreducible chain on C_r by (7.42). Thus $(d\pi_i : i \in C_r)$ is its only invariant distribution, that is,

$$P(Y_{nd+r} = j) = \begin{cases} d\pi_j & \text{if } j \in C_r, \\ 0 & \text{otherwise}. \end{cases} \tag{7.44}$$

As a first step, we shall show that as $n \to \infty$,

$$\frac{1}{\sqrt{n}} \sum_{i=1}^{n} f(Y_i) \Rightarrow Z. \tag{7.45}$$

First note that

$$E|f(Y_1)| = d \sum_{j \in C_1} \pi_j |f(j)|$$

$$\leq d \sum_{j=1}^{k} \pi_j |f(j)|$$

$$\leq d \sum_{j=1}^{k} \pi_j (1 + f(j)^2) < \infty \ (\text{by } (7.39)).$$

Hence

$$\infty > \sum_{j=1}^{k} \pi_j |f(j)| = \sum_{j=1}^{k} \sum_{i=1}^{k} \pi_i p_{ij} |f(j)| \ (\text{from } (7.37))$$

$$= \sum_{i=1}^{k} \pi_i \sum_{j=1}^{k} p_{ij} |f(j)|.$$

We now verify that $f(Y_n)$ has finite mean for all n. For $n \geq 2$,

$$E|f(Y_n)| = \sum_{i=1}^{k} \sum_{j=1}^{k} |f(j)| p_{ij} P(X_{n-1} = i)$$

$$\leq d \sum_{i=1}^{k} \sum_{j=1}^{k} |f(j)| p_{ij} \pi_i \ (\text{using } (7.44))$$

$$= d \sum_{i=1}^{k} \pi_i \sum_{j=1}^{k} |f(j)| p_{ij} < \infty.$$

Now (7.38) says that

$$\mathrm{E}(f(Y_n)|Y_{n-1}) = 0, \; n \geq 2.$$

The Markov property implies that $(M_n, \mathcal{G}_n)$ is a martingale where

$$M_n = \sum_{i=1}^{n} f(Y_i), \; \mathcal{G}_n = \sigma(Y_1, \ldots, Y_n) \text{ for } n \geq 1.$$

In view of (7.39), we can mimic the proof of finiteness of the expectation, to show that $\mathrm{V}(f(Y_n)) < \infty$.

Use the Markov property once again to write

$$\mathrm{E}\left((M_{n+1} - M_n)^2 | \mathcal{G}_n\right) = g(Y_n), \; n \geq 1,$$

where

$$g(i) = \sum_{j=1}^{k} p_{ij} f(j)^2, \; i \in \mathcal{S}.$$

Note that

$$\sum_{i=1}^{k} \pi_i g(i) = \sum_{i=1}^{k} \sum_{j=1}^{k} \pi_i p_{ij} f(j)^2$$

$$= \sum_{j=1}^{k} f(j)^2 \pi_j \text{ (using (7.37))}$$

$$= \sigma^2 < \infty. \tag{7.46}$$

We shall now show that

$$\lim_{n \to \infty} \frac{1}{n} \sum_{i=1}^{n} g(Y_i) = \sum_{j=1}^{k} \pi_j g(j) \text{ in } L^1. \tag{7.47}$$

Fix $\varepsilon > 0$. Using (7.46), get $N \in \mathbb{N}$, $N \leq k$ such that $\sum_{j=N+1}^{k} \pi_j |g(j)| \leq \varepsilon$. Define the truncated version of g as $g'(i) = g(i)\mathbf{1}(i \leq N)$ for $i \in \mathcal{S}$. Theorem 1.10.2 of Norris

7.5 Markov Chains

(1998) shows that $\frac{1}{n}\sum_{i=1}^{n} g'(Y_i) \to \sum_{j=1}^{k} \pi_j g'(j)$, almost surely. Use DCT to rewrite the above as

$$\frac{1}{n}\sum_{i=1}^{n} g'(Y_i) \to \sum_{j=1}^{N} \pi_j g(j) \text{ in } L^1. \qquad (7.48)$$

For any $i \geq 1$,

$$\begin{aligned}
\mathrm{E}\left|g(Y_i) - g'(Y_i)\right| &= \sum_{j=N+1}^{k} |g(j)| P(Y_i = j) \\
&\leq d \sum_{j=N+1}^{k} \pi_j |g(j)| \text{ (using (7.44))}.
\end{aligned}$$

The choice of N shows that

$$\mathrm{E}\left|g(Y_i) - g'(Y_i)\right| \leq d\varepsilon.$$

Combine this with (7.48) and use the choice of N again to get

$$\limsup_{n \to \infty} \mathrm{E}\left|\frac{1}{n}\sum_{i=1}^{n} g(Y_i) - \sum_{j=1}^{k} \pi_j g(j)\right| \leq (d+1)\varepsilon.$$

Arbitrariness of ε proves (7.47). Thus

$$\frac{1}{n}\sum_{i=1}^{n} \mathrm{E}\left((M_{i+1} - M_i)^2 | \mathcal{G}_i\right) \to \sigma^2 \text{ in } L^1.$$

For a fixed $\varepsilon > 0$, which is unrelated to the ε chosen in the above paragraph, write

$$\begin{aligned}
\frac{1}{n}\sum_{i=1}^{n} &\mathrm{E}\left((M_i - M_{i-1})^2 \mathbf{1}_{\{|M_i - M_{i-1}| \geq \varepsilon\sqrt{n}\}}\right) \\
&\leq \frac{1}{n} \sum_{i=1}^{d\lceil n/d \rceil} \mathrm{E}\left(f(Y_i)^2 \mathbf{1}_{\{|f(Y_i)| \geq \varepsilon\sqrt{n}\}}\right) \\
&= \frac{1}{n} \lceil n/d \rceil \sum_{r=1}^{d} \mathrm{E}\left(f(Y_r)^2 \mathbf{1}_{\{|f(Y_r)| \geq \varepsilon\sqrt{n}\}}\right) \to 0 \text{ as } n \to \infty.
\end{aligned}$$

Here $\lceil x \rceil = $ smallest integer $\geq x$, and equality in the penultimate line follows from (7.43).

Theorem 7.2.1 then implies that $n^{-1/2}(M_n - M_0) \Rightarrow Z$, from which (7.45) follows.

Recall that (Y_n) was chosen to be independent of (X_n). We finish the proof with a *coupling argument*. Define $W_n := (X_n, Y_n)$, $n \geq 1$. Then $(W_n : n \geq 1)$ is a Markov chain on $\mathcal{S} \times \mathcal{S}$ with transition probabilities $(\tilde{p}_{(i,j),(i',j')} : i, j, i', j' \in \mathcal{S})$ defined by $\tilde{p}_{(i,j),(i',j')} := p_{ii'} p_{jj'}$.

For $n \geq 0$, $X_{nd+1}, Y_{nd+1} \in C_1$, that is, $W_{nd+1} \in C_1 \times C_1$. For $i, j, i', j' \in C_1$,

$$P\left(W_{nd+1} = (i', j') | W_1 = (i, j)\right) = p_{ii'}^{(nd)} p_{jj'}^{(nd)}.$$

It follows from (7.42) that the right side above is strictly positive for n large. Hence, $(W_{nd+1} : n \geq 0)$ is an irreducible chain on $C_1 \times C_1$. Further, $(d^2 \pi_i \pi_j : (i, j) \in C_1 \times C_1)$ is an invariant distribution for $(W_{nd+1} : n \geq 0)$, which is therefore positive recurrent, and, in particular, recurrent. Recall that $a \in C_1$, and hence

$$\inf\{n \geq 0 : W_{nd+1} = (a, a)\} < \infty \text{ almost surely}.$$

Let

$$T = \inf\{n \geq 1 : W_n = (a, a)\},$$

which is thus finite almost surely. For $n \geq 1$, let

$$X'_n = \begin{cases} X_n & \text{if } n \leq T, \\ Y_n & \text{if } n > T. \end{cases}$$

Since T is almost surely finite, $\frac{1}{\sqrt{n}} \sum_{i=1}^{n} \left(f(X'_i) - f(Y_i)\right) \to 0$ a.s., which in conjunction with (7.45) shows that $\frac{1}{\sqrt{n}} \sum_{i=1}^{n} f(X'_i) \Rightarrow Z$.

The above would establish (7.41) if it can be shown that

$$(X_n : n \geq 1) \stackrel{d}{=} (X'_n : n \geq 1). \tag{7.49}$$

For this we use the *strong Markov property* (see Theorem 1.4.2 of Norris (1998)). This property implies that $(W_{T+n} : n \geq 0)$ is a Markov chain with initial distribution $\delta_{(a,a)}$, independent of $(W_0, \ldots, W_T)$. Defining

$$W'_n = (Y_{T+n}, X_{T+n}), n \geq 0,$$

symmetry implies $(W'_n : n \geq 0) \stackrel{d}{=} (W_{T+n} : n \geq 0)$. Hence,

$$\left((W_0, \ldots, W_T), (W_{T+n} : n \geq 1)\right) \stackrel{d}{=} \left((W_0, \ldots, W_T), (W'_n : n \geq 1)\right).$$

Using the first coordinate from above, we obtain

$$(X_1, X_2, \ldots) \stackrel{d}{=} (X_1, \ldots, X_T, Y_{T+1}, Y_{T+2}, \ldots).$$

As the right side above is the same as $(X'_1, X'_2, \ldots)$, (7.49) follows, which in turn establishes (7.41). This completes the proof. □

References

B. Friedman, A simple urn model. Commun. Pure Appl. Math. **2**(1), 59–70 (1949)

D.A. Freedman, Bernard Friedman's Urn. Ann. Math. Stat. **36**(3), 956–970 (1965)

P. Hall, C.C. Heyde, *Martingale Limit Theory and its Application* (Academic, Harcourt Brace Jovanovich, Publishers, New York-London, 1980)

P. Lévy, *Théorie de l'Addition des Variables Aléatoires* (Gauthier-Villars, Paris, 1937)

J.W. Lindeberg, Eine neue herleitung des exponentialgesetzes in der wahrscheinlichkeitsrechnung. Mathematische Zeitschrift **15**, 211–225 (1922)

J.R. Norris, *Markov Chains, Cambridge Series in Statistical and Probabilistic Mathematics*, vol. 2. (Cambridge University Press, Cambridge, 1998). Reprint of 1997 original

Chapter 8
Additional Topics

This chapter covers some additional topics that the reader may find interesting. It includes a forward representation for U-statistics, the weak and strong Azuma–Hoeffding exponential inequalities, conditional versions of the Borel–Cantelli lemma, and Kolmogorov's three-series theorem, SLLN for martingales, and finally the Kesten–Stigum theorem for a simple branching process.

8.1 U-Statistics, Forward Martingale Representation

We have proved the U-statistics SLLN using a reverse martingale representation. A proof can also be constructed using a forward martingale representation, which shows that any U-statistics with an mth order kernel is a weighted sum of m martingales, where the weights depend on n and m. We give this representation which is due to Hoeffding (1961).

Theorem 8.1.1 (U-statistics, martingale representation) *Suppose $\{X_i\}$ are iid and U_n is a U-statistics with kernel h of order n which has finite mean μ. Then we can write*

$$U_n - \mu = \sum_{c=1}^{m} \binom{m}{c}\binom{n}{c}^{-1} S_{cn}$$

where, for each $1 \leq c \leq m$, $\left(S_{cn}, \sigma(X_1, \ldots, X_n)\right)_{n \geq m}$ is a zero mean martingale. ◆

Proof Define kernels of order c, $1 \leq c \leq m$ as

$$h_c(x_1, \ldots, x_c) := \mathrm{E}\big(h(X_1, \ldots, X_m) | X_1 = x_1, \ldots, X_c = x_c\big).$$

Clearly, $h_c(X_1, \ldots, X_c)$ is a version of the conditional expectation of $h(X_1, \ldots, X_m)$ given $\sigma(X_1, \ldots, X_c)$.

Further, by the tower property (Theorem 3.1.2),

$$h_c(x_1, \ldots, x_c) = \mathrm{E}\bigl(h_{c+1}(x_1, \ldots, x_c, X_{c+1})\bigr).$$

For ease of computations, define the centered version of these functions as

$$\tilde{h} := h - \mu \quad \tilde{h}_c := h_c - \mu, \ 1 \leq c \leq m. \tag{8.1}$$

We define a sequence of functions that "orthogonalize" the sequence $\{h_c\}$, $1 \leq c \leq m$. Let

$$\begin{aligned}
g_1(x_1) &:= \tilde{h}_1(x_1), \\
g_2(x_1, x_2) &:= \tilde{h}_2(x_1, x_2) - g_1(x_1) - g_2(x_2), \\
g_3(x_1, x_2, x_3) &:= \tilde{h}_3(x_1, x_2, x_3) - \sum_{i=1}^{3} g_1(x_i) - \sum_{1 \leq i_1 < i_2 \leq 3}^{3} g_2(x_{i_1}, x_{i_2}), \\
\cdots &= \cdots \\
g_m(x_1, \ldots, x_m) &:= \tilde{h}(x_1, \ldots, x_m) - \sum_{i=1}^{m} g_1(x_i) - \sum_{1 \leq i_1 < i_2 \leq m} g_2(x_{i_1}, x_{i_2}) \\
&\quad - \cdots - \sum_{1 \leq i_1 < \cdots < i_{m-1} \leq m} g_{m-1}(x_{i_1}, \ldots x_{i_{m-1}}).
\end{aligned}$$

Note that each g_c is a kernel of order c. Moreover, from the way we have constructed these functions,

$$\mathrm{E}(g_1(X_1)) = \mathrm{E}(g_2(x_1, X_2)) = \cdots = \mathrm{E}(g_m(x_1, \ldots, x_{m-1}, X_m)) = 0.$$

Now for any $n \geq m$, let

$$\begin{aligned}
S_{c,n} &:= \sum_{1 \leq i_1 < \cdots < i_c \leq n} g_c(X_{i_1}, \ldots, X_{i_c}), \\
S_n &:= \sum_{1 \leq i_1 < \cdots < i_m \leq n} \tilde{h}(X_{i_1}, \ldots, X_{i_m}).
\end{aligned}$$

Then, using the relation between $\{g_c\}$, $1 \leq c \leq m$ and $\tilde{h}$, we have

$$\begin{aligned}
S_n &= S_{m,n} + \sum_{c=1}^{m-1} \sum_{1 \leq i_1 < \cdots < i_m \leq n} \sum_{1 \leq j_1 < \cdots < j_c \leq m} g_c(X_{i_{j_1}}, \ldots, X_{i_{j_c}}) \\
&= \sum_{c=1}^{m} \binom{n}{c}^{-1} \binom{n}{m} \binom{m}{c} S_{cn},
\end{aligned}$$

8.2 Azuma–Hoeffding Inequality

where the last equality follows from counting the number of times each $g_c(X_{i_{j_1}}, \ldots, X_{i_{j_c}})$ appears in the sum. To show the martingale property,

$$\begin{aligned}
E(S_{c,n+1}|X_1, \ldots, X_n)) &= \sum_{1 \leq i_1 < \cdots < i_c \leq n+1} E\Big(g_c(X_{i_1}, \ldots, X_{i_c})|X_1, \ldots, X_n\Big) \\
&= \sum_{1 \leq i_1 < \cdots < i_c \leq n} g_c(X_{i_1}, \ldots, X_{i_c}) \\
&\quad + \sum_{1 \leq i_1 < \cdots < i_c = n+1} E\Big(g_c(X_{i_1}, \ldots, X_{i_c})|X_1, \ldots, X_n\Big) \\
&= \sum_{1 \leq i_1 < \cdots < i_c \leq n} g_c(X_{i_1}, \ldots, X_{i_c}) = S_{c,n}.
\end{aligned}$$

The second term after + sign vanished by upon using the definition of $g_c(\cdot)$. This proves the result since $U_n - \mu = \binom{n}{m}^{-1} S_n$. □

Exercise 8.1.1 (*U*-statistics CLT) Suppose U_n is a *U*-statistics based on iid observations $\{X_n\}$ and kernel h of order m. Let $\tilde{h}_1$ be as in (8.1) of Theorem 8.1.1 and suppose $\sigma^2 := E(\tilde{h}_1(X_1)^2) < \infty$. Show that as $n \to \infty$, $n^{1/2}(U_n - E(h(X_1, \ldots X_m)) \Rightarrow N(0, m^2\sigma^2)$.

8.2 Azuma–Hoeffding Inequality

The upper bound in Markov inequality, Theorem 1.6.1(i), decreases polynomially in the argument. In any context, it is an interesting question as to whether there are bounds that decrease more sharply. It turns out that in many cases, bounds can be obtained which decrease exponentially, and are often technically extremely useful in advanced probability. Historically, the earliest such inequalities were proved by S. N. Bernstein in the 1920s and 1930s for bounded random variables (Bernstein (1924, 1927, 1937)), and later by other authors. These are collectively known as Bernstein inequalities. Over the years there have been many extensions and generalizations to different probability models and these now go by the name of *exponential inequalities* and *concentration inequalities*. We shall state and prove an exponential inequality for martingales which satisfy a relatively strong condition. For other strong exponential inequalities especially for martingales and applications in different areas, the reader is referred to McDiarmid (1989), Bercu et al. (2015).

We first prove an easier version of the Azuma–Hoeffding inequality.

Theorem 8.2.1 (Weak Azuma–Hoeffding inequality) *Let $(X_n, \mathcal{G}_n)$ be a martingale. Suppose that for some positive constants $\{\sigma_k\}$,*

$$|X_n - X_{n-1}| \leq \sigma_n, \; n \geq 1,$$

Then for all $t \geq 0$ and $n \geq 1$,

$$P(|X_n - X_0| \geq t) \leq 2\exp\left(\frac{-t^2}{2\sum_{k=1}^{n}\sigma_k^2}\right).$$

♦

We first state an exercise which we will use in the proof.

Exercise 8.2.1 For all $n \geq 1$ and $\sigma^2, t > 0$,

$$\min_{\theta>0} \exp\left(-\theta t + \frac{\theta^2}{2}\sigma^2\right) = \exp\left(-\frac{t^2}{2\sigma^2}\right).$$

Proof of Theorem 8.2.1 Since $\{-X_n\}$ is also a martingale, it is enough to prove the one-sided inequality,

$$P(X_n - X_0 \geq t) \leq \exp\left(\frac{-t^2}{2\sum_{k=1}^{n}\sigma_k^2}\right).$$

Let $\theta > 0$. The value of θ will be chosen later. Using Markov inequality we have

$$\begin{aligned}P(X_n - X_0 \geq t) &= P(e^{\theta(X_n - X_0)} \geq e^{\theta t})\\ &\leq e^{-\theta t}\mathrm{E}(e^{\theta(X_n - X_0)})\\ &= e^{\theta t}\mathrm{E}(e^{\theta(X_{n-1} - X_0)}e^{\theta\sigma_n D_n}),\end{aligned} \quad (8.2)$$

where $D_n = (X_n - X_{n-1})/\sigma_n$. Note that $|D_n| \leq 1$, and $\mathrm{E}(D_n|\mathcal{G}_{n-1}) = 0$.
So we have

$$P(X_n - X_0 \geq t) \leq e^{-\theta t}\mathrm{E}\left(e^{\theta(X_{n-1} - X_0)}\mathrm{E}\left(e^{\theta\sigma_n D_n}|\mathcal{G}_{n-1}\right)\right). \quad (8.3)$$

For every $t > 0$, the function $y \mapsto e^{ty}$ is a convex function of y. For $|x| \leq 1$, write tx as,

$$tx = \frac{1-x}{2}(-t) + \frac{1+x}{2}t \quad \text{(a convex combination)}.$$

Hence,

$$e^{tx} \leq \frac{1-x}{2}e^{-t} + \frac{1+x}{2}e^t.$$

This inequality gives us

$$e^{\theta\sigma_n D_n} \leq \frac{1 - D_n}{2}e^{-\theta\sigma_n} + \frac{1 + D_n}{2}e^{\theta\sigma_n}.$$

Hence

8.2 Azuma–Hoeffding Inequality

$$E(e^{\theta \sigma_n D_n}|\mathcal{G}_{n-1}) \leq \frac{1}{2}(e^{-\theta \sigma_n} + e^{\theta \sigma_n})$$

$$= \sum_{i=0}^{\infty} \frac{(\theta \sigma_n)^{2i}}{2i!}$$

$$\leq \sum_{i=0}^{\infty} \frac{(\theta \sigma_n)^{2i}}{i! 2^i} = e^{\frac{1}{2}\theta^2 \sigma_n^2}, \qquad (8.4)$$

where the inequality in (8.4) follows from the observation

$$\frac{(2n)!}{2^n n!} = \text{Number of pair-partitions of } \{1, \ldots, 2n\} \geq 1.$$

So plugging this estimate into (8.3), we obtain

$$P(X_n - X_0 \geq t) \leq \exp\left(-\theta t + \frac{\theta^2 \sigma_n^2}{2}\right) E\left(e^{\theta(X_{n-1} - X_0)}\right).$$

Now repeating the above argument we get

$$P(X_n - X_0 \geq t) \leq \exp\left(-\theta t + \frac{\theta^2 \sum_{i=1}^{n} \sigma_i^2}{2}\right).$$

Using Exercise 8.2.1, the result follows. $\square$

The *chromatic number* of a graph is the minimum number of colors required to color each vertex such that no two adjacent vertices have the same color. For Erdős-Rényi random graph on n vertices $v_1, v_2, \ldots, v_n$, there is an edge between each pair of vertices independently with probability p. Hence its chromatic number is a random variable.

Corollary 8.2.1 (Chromatic number of Erdős-Rényi random graph) *Let ξ be the chromatic number of the Erdős-Rényi random graph G with n vertices $v_1, \ldots, v_n$. Then,*

$$P(|\xi - E(\xi)| > x) \leq 2 \exp\left(-\frac{x^2}{2n}\right), \text{ for all } x > 0.$$

♦

Proof Let $\mathcal{G}_0 = \{\phi, \Omega\}$, $\mathcal{G}_i$ be the σ-field generated by the (indicator of) edges between $v_1, \ldots, v_i$. Then $Z_i = E(\xi|\mathcal{G}_i)$ is a martingale. Also note that, $Z_n = E(\xi|\mathcal{G}_n) = \xi$ and $Z_0 = E(\xi|\mathcal{G}_0) = E(\xi)$.

Let $G(j)$ be the random graph obtained by erasing the jth vertex, and all edges connected to it. Let $\xi(j)$ be the chromatic number of $G(j)$. We immediately see that

$$\xi(j) \leq \xi \leq \xi(j) + 1,$$
$$E[\xi(j)|\mathcal{G}_{j-1}] = E[\xi(j)|\mathcal{G}_j],$$

$$\begin{aligned} Z_i &\leq \mathrm{E}(\xi(i)|\mathcal{G}_i)+1 \\ &= \mathrm{E}[\xi(i)|\mathcal{G}_{i-1}]+1 \\ &\leq \mathrm{E}(\xi|\mathcal{G}_{i-1})+1 = Z_{i-1}+1. \end{aligned}$$

As a consequence,
$$|Z_i - Z_{i-1}| \leq 1.$$

The result now follows by applying Theorem 8.2.1. □

Theorem 8.2.2 (Strong Azuma–Hoeffding inequality) *Suppose that $(X_n, \mathcal{G}_n)$ is a martingale with $X_0 = 0$. Further,*

$$|X_n - X_{n-1}| \leq \sigma_n, \, n \geq 1,$$

for some strictly positive $\mathcal{G}_0$-measurable random variable σ_n. Then for all $t \geq 0$ and $n \geq 1$,

$$P(X_n \geq t|\mathcal{G}_0) \leq \exp\left(\frac{-t^2}{2\sum_{k=1}^{n}\sigma_k^2}\right).$$

♦

The following elementary lemma will be used in the proof of Theorem 8.2.2.

Lemma 8.2.1 *Suppose X is a random variable with values in the interval $[0, 1]$ and $\mathrm{E}(X) = 1/2$. Let $\phi : [0, 1] \to \mathbb{R}$ be a convex function. Then $\mathrm{E}(\phi(X)) \leq \frac{\phi(0)+\phi(1)}{2}$.*

♦

Proof (*On an appropriate probability space*) Let Z be a random variable, independent of X, and distributed uniformly on the interval $[0, 1]$. Then

$$\phi(X) = \phi\left(\mathrm{E}(\mathbf{1}_{\{Z \leq X\}}|\sigma(X))\right) \leq \mathrm{E}\left(\phi\left(\mathbf{1}_{\{Z \leq X\}}\right)|\sigma(X)\right),$$

where the last inequality follows from Jensen's inequality Theorem 3.3.1. Hence,

$$\begin{aligned} \mathrm{E}(\phi(X)) &\leq \mathrm{E}\phi\left(\mathbf{1}_{\{Z \leq X\}}\right) \\ &= \frac{\phi(0)+\phi(1)}{2}, \end{aligned}$$

where the equality holds since $\mathbf{1}_{\{Z \leq X\}}$ takes values 0, 1, and

$$P(Z \leq X) = \mathrm{E}P(Z \leq X|\sigma(X)) = \mathrm{E}(X) = \frac{1}{2}.$$

Hence the proof is complete. □

8.2 Azuma–Hoeffding Inequality

Proof of Theorem 8.2.2 Note that for all $t, \theta > 0$,

$$P(X_n \geq t | \mathcal{G}_0) = P(e^{\theta X_n} \geq e^{\theta t} | \mathcal{G}_0) \leq e^{-\theta t} E(e^{\theta X_n} | \mathcal{G}_0). \tag{8.5}$$

If we prove the following statement then the result follows from Exercise 8.2.1.

For all $n \geq 1$ and $\theta > 0$,

$$E(e^{\theta X_n} | \mathcal{G}_0) \leq \exp\left(\frac{1}{2}\theta^2 \sum_{j=1}^{n} \sigma_j^2\right). \tag{8.6}$$

We shall use induction on n to prove (8.6). Let $n = 1$, Recall that $|X_1| \leq \sigma_1$. Define

$$Y := \frac{X_1 + \sigma_1}{2\sigma_1},$$

and $\phi : \mathbb{R}^2 \to \mathbb{R}$ by

$$\phi(y, z) := e^{\theta(2yz - z)} \mathbf{1}_{\{y \in [0,1]\}}, \quad (y, z) \in \mathbb{R}^2.$$

Let μ be the RCD of Y given $\mathcal{G}_0$. Since σ_1 is $\mathcal{G}_0$-measurable, by Exercise 3.5.22, we have

$$\left(E(e^{\theta X_1} | \mathcal{G}_0)\right)(\omega) = \left(E\left(\phi(Y, \sigma_1) | \mathcal{G}_0\right)\right)(\omega)$$

$$= \int_0^1 \phi(y, \sigma_1(\omega)) \mu(dy, \omega) \text{ for all } \omega \in \Omega.$$

Note that σ_1 is $\mathcal{G}_0$-measurable and $E(X_1 | \mathcal{G}_0) = 0$. Hence $E(Y | \mathcal{G}_0) = \frac{1}{2}$. Hence observing that ϕ is convex in the first variable on $[0, 1]$, by Lemma 8.2.1, we have

$$\int_0^1 \phi(y, \sigma_1(\omega)) \mu(dy, \omega) \leq \frac{1}{2}[\phi(0, \sigma_1(\omega)) + \phi(1, \sigma_1(\omega))]$$

$$= \frac{1}{2}\left(e^{-\theta \sigma_1(\omega)} + e^{\theta \sigma_1(\omega)}\right).$$

Thus,

$$E(e^{\theta X_1} | \mathcal{G}_0) \leq \frac{1}{2}\left(e^{-\theta \sigma_1} + e^{\theta \sigma_1}\right)$$

$$\leq e^{\theta^2 \sigma_1^2 / 2} \text{(as in (8.4))}. \tag{8.7}$$

Thus, (8.6) is true for $n = 1$. Assume that (8.6) holds for $n - 1$ for some $n \geq 2$. Write

$$E\left(e^{\theta X_n} | \mathcal{G}_0\right) = E\left(e^{\theta X_{n-1}} E\left(e^{\theta(X_n - X_{n-1})} | \mathcal{G}_{n-1}\right) | \mathcal{G}_0\right).$$

Define
$$Z := \frac{1}{2\sigma_n}(X_n - X_{n-1} + \sigma_n),$$

and $\psi : [0,1] \times \mathbb{R} \to \mathbb{R}$ by
$$\psi(z,v) := e^{\theta(2vz-v)}.$$

The hypothesis $|X_n - X_{n-1}| \leq \sigma_n$ implies that $0 \leq Z \leq 1$. Using the fact that σ_n is measurable with respect to $\mathcal{G}_0$ and hence $\mathcal{G}_n$, and applying Exercise 3.5.22, we get
$$\mathrm{E}\!\left(e^{\theta(X_n-X_{n-1})}|\mathcal{G}_{n-1}\right) = \int_0^1 \psi(z,\sigma_n) P(Z \in dz|\mathcal{G}_{n-1}),$$

where $P(Z \in \cdot | \mathcal{G}_{n-1})$ is the RCD of Z given $\mathcal{G}_{n-1}$. Using Lemma 8.2.1 and the fact that almost surely,
$$\int_0^1 z P(Z \in dz|\mathcal{G}_{n-1}) = \mathrm{E}(Z|\mathcal{G}_{n-1}) = \frac{1}{2},$$

it follows that
$$\int_0^1 \psi(z,\sigma_n) P(Z \in dz|\mathcal{G}_{n-1}) \leq \frac{1}{2}[\psi(0,\sigma_n) + \psi(1,\sigma_n)] \quad \text{almost surely}.$$

Arguments similar to (8.4) shows that
$$\mathrm{E}\!\left(e^{\theta(X_n-X_{n-1})}|\mathcal{G}_{n-1}\right) \leq e^{\theta^2 \sigma_n^2/2}.$$

Therefore,
$$\mathrm{E}\!\left(e^{\theta X_n}|\mathcal{G}_0\right) \leq e^{\theta^2 \sigma_n^2/2} \mathrm{E}\!\left(e^{\theta X_{n-1}}|\mathcal{G}_0\right)$$
$$\leq \exp\!\left(\frac{1}{2}\theta^2 \sum_{j=1}^n \sigma_j^2\right) \quad \text{(using the induction hypothesis)}.$$

Then (8.6) follows. This completes the proof of Theorem 8.2.2 in view of (8.5) and Exercise 8.2.1. $\square$

8.3 Weak Law for Martingales

Theorem 8.3.1 (Weak law for martingales) *Suppose $(X_n, \mathcal{G}_n)_{n \geq 0}$ is an adapted sequence. Let $S_n = \sum_{i=1}^n X_i$. For some sequence $b_n \uparrow \infty$, let $X_{ni} = X_i \mathbf{1}_{\{|X_i| \leq b_n\}}$. Suppose the following three conditions hold:*

(a) $\sum_{i=1}^{n} P\{|X_i| > b_n\} \to 0.$

(b) $b_n^{-1} \sum_{i=1}^{n} E(X_{ni}|\mathcal{G}_{i-1}) \to 0$ *in probability.*

(c) $b_n^{-2} \sum_{i=1}^{n} \left[E(X_{ni}^2) - E\left((E(X_{ni}|\mathcal{G}_{i-1}))^2\right) \right] \to 0$ *in probability.*

Then $b_n^{-1} S_n \to 0$ *in probability.* ♦

Incidentally, if $\{X_n\}$ are independent, then conditions (a), (b), and (c) are also necessary for the convergence (see Loève (1977)). Proof of this is beyond the scope of this book.

Proof Let $S_{nn} = \sum_{i=1}^{n} X_{ni}$. Then

$$P\{S_{nn} \neq S_n\} \leq \sum_{i=1}^{n} P\{X_{ni} \neq X_i\}$$
$$= \sum_{i=1}^{n} P\{|X_i| > b_n\} \to 0, \quad \text{(by Condition (a))}.$$

Hence it is enough to prove that $\dfrac{S_{nn}}{b_n} \to 0$ in probability.

Now using Condition (b), it is enough to prove that

$$\frac{1}{b_n} \sum_{i=1}^{n} \left(X_{ni} - E(X_{ni}|\mathcal{G}_{i-1}) \right) \to 0 \text{ in probability.}$$

In view of Condition (c), the result follows by using Chebyshev's or Markov's inequality. □

8.4 Extended Borel–Cantelli Lemma

Many extensions of the Borel–Cantelli lemma have appeared to cater to different situations. We establish an extension using martingales.

We need the following result.

Lemma 8.4.1 *Suppose that* $(X_n, \mathcal{G}_n)$ *is an adapted sequence and that* $(S_n = \sum_{i=1}^{n} X_i, \mathcal{G}_n)_{n \geq 1}$ *is a martingale. Assume that* $E(X_n) = 0$ *for all* n, *and* $E(\sup_n |X_n|) < \infty$. *Then* $\limsup S_n = \infty$ *and* $\liminf S_n = -\infty$, *on the set* $A = \{\omega : \sum_{n=1}^{\infty} X_n(\omega) \text{ diverges}\}$. ♦

Proof For any $a \in \mathbb{R}$, define

$$\tau_a := \begin{cases} \inf\{n : S_n > a\} & \text{if there is such an } n, \\ \infty & \text{if there is no such } n. \end{cases} \tag{8.8}$$

Then $\{\tau_{a \wedge n}\}$ is a sequence of bounded non-decreasing $\mathcal{G}_{\tau_{a \wedge n}}$ stop times. By Doob's optional sampling Theorem 4.3.2, $(S_{\tau_{a \wedge n}}, \mathcal{G}_{\tau_{a \wedge n}})_{n \geq 1}$ is a zero mean martingale and so, $E(|S_{\tau_{a \wedge n}}|) = 2E(S^+_{\tau_{a \wedge n}})$. Now

$$S^+_{\tau_{a \wedge n}} \leq S_{\tau_{a \wedge (n-1)}} + X^+_{\tau_{a \wedge n}} \leq a + \sup X^+_n.$$

Since $E(\sup_n |X_n|) < \infty$, we have $E(|S_{\tau_{a \wedge n}}|)$ is a bounded sequence. By Martingale convergence Theorem 5.2.1, as $n \to \infty$, $S_{\tau_{a \wedge n}}$ converges almost surely to a finite limit.

As a result, $\lim_{n \to \infty} S_n$ exists and is almost surely finite on the set $B_a := \{\omega : \sup_n S_n(\omega) \leq a\}$.

Let $a \to \infty$ (through rationals) to infer that $\lim_{n \to \infty} S_n$ exists and is a.s. finite on $B = \{\omega : \sup_n S_n(\omega) < \infty\} = \{\omega : \limsup_n S_n(\omega) < \infty\}$. As a consequence, $\limsup_n S_n(\omega) = \infty$ on the set $\{\omega : S_n(\omega) \text{ diverges}\}$.

The lim inf claim follows by using the above argument on $\{-S_n\}$. $\square$

Theorem 8.4.1 (Extended Borel–Cantelli lemma) *(a) Suppose that $(Z_n, \mathcal{G}_n)$ is an adapted sequence, $0 \leq Z_n \leq 1$. Then $\sum_{n=1}^\infty Z_n$ converges a.s. if and only if $\sum_{n=1}^\infty E(Z_n|\mathcal{G}_{n-1}) < \infty$ a.s.*
(b) Hence, if $(Z_n, \mathcal{G}_n)$ is an adapted sequence and $A_n \in \sigma(Z_1, \ldots, Z_n)$, then the sets $\limsup_{n \to \infty} A_n$ and $\{\omega : \sum_{n=1}^\infty P(A_n|\mathcal{G}_{n-1}) = \infty\}$ are equal almost surely. ◆

Proof (a) Define for $n \geq 1$,

$$\mathcal{G}_0 := \{\emptyset, \Omega\}, \ X_n := Z_n - E(Z_n|\mathcal{G}_{n-1}), \text{ and } S_n := \sum_{i=1}^n X_i.$$

Then $(S_n, \mathcal{G}_n)_{n \geq 1}$ is a martingale with uniformly bounded differences. Further, $\limsup S_n \leq \sum_{i=1}^\infty Z_i$. By Lemma 8.4.1, $S := \lim S_n$ exists and is finite almost surely on the set $B := \{\omega : \sum_{i=1}^\infty Z_i(\omega) < \infty\}$. Further

$$\sum_{i=1}^\infty E(Z_i|\mathcal{G}_{i-1}) = \sum_{i=1}^\infty Z_i - S$$

converges a.s. on B.

Similarly, using $\liminf S_n \geq -\sum_{i=1}^\infty E(Z_i|\mathcal{G}_{i-1})$, we can show that

$$\sum_{i=1}^\infty Z_i(\omega) < \infty \text{ almost surely on } \{\omega : \sum_{i=1}^\infty E(Z_i|\mathcal{G}_{i-1}) < \infty\}.$$

This proves (a).

(b) This follows immediately from Part (a) once we observe that

$$\limsup A_n = \{\omega : \sum_{i=1}^{\infty} \mathbf{1}_{A_n}(\omega) = \infty\} \text{ and } P(A_{n+1}|\mathcal{G}_n) = \mathrm{E}(\mathbf{1}_{A_{n+1}}|\mathcal{G}_n).$$

□

8.5 Three-Series Theorem

In a first course on probability, the celebrated three-series theorem of Kolmogorov is proved for independent random variables. We prove a martingale version due to Brown (1971). We shall need a generalization of the convergence for L^2 bounded martingales.

Theorem 8.5.1 *Suppose $(X_n, \mathcal{G}_n)_{n \geq 0}$ is an adapted sequence, such that $(S_n = \sum_{i=1}^{n} X_i, \mathcal{G}_n)_{n \geq 0}$ is a zero mean martingale. Then S_n converges a.s. on the set $\{\omega : \sum_{i=1}^{\infty} \mathrm{E}(X_i^2|\mathcal{G}_{i-1}) < \infty\}$.* ◆

Proof Define $\mathcal{G}_0 := \{\emptyset, \Omega\}$. Fix $K > 0$ and define the stop time,

$$\tau := \begin{cases} \inf\{n : \sum_{i=1}^{n} \mathrm{E}(X_i^2|\mathcal{G}_{i-1}) > K\} & \text{if there is such an } n, \\ \infty & \text{if there is no such } n. \end{cases} \quad (8.9)$$

Then $(S_{\tau \wedge n}, \mathcal{G}_{\tau \wedge n})_{n \geq 1}$ is a zero mean martingale. Moreover, note that $\mathbf{1}_{\{\tau \geq i\}}$ is $\mathcal{G}_{i-1}$-measurable. We can write $S_{\tau \wedge n} = \sum_{i=1}^{n} \mathbf{1}_{\{\tau \geq i\}} X_i$.

Hence,

$$\mathrm{E}(S_{\tau \wedge n}^2) = \mathrm{E}\Big(\sum_{i=1}^{n} \mathbf{1}_{\{\tau \geq i\}} X_i^2\Big)$$

$$= \mathrm{E}\Big(\sum_{i=1}^{n} \mathrm{E}\Big(\mathbf{1}_{\{\tau \geq i\}} X_i^2 | \mathcal{G}_{i-1}\Big)\Big)$$

$$= \mathrm{E}\Big(\sum_{i=1}^{\tau \wedge n} \mathrm{E}(X_i^2 | \mathcal{G}_{i-1})\Big)$$

$$\leq K.$$

Hence $\{S_{\tau \wedge n}\}$ is L^2 bounded. By Martingale convergence Theorem 5.3.1, $S_{\tau \wedge n}$ converges almost surely. Hence S_n converges almost surely on the set $\{\omega : \tau(\omega) = \infty\} = \{\omega : \sum_{i=1}^{n} \mathrm{E}(X_i^2|\mathcal{G}_{i-1}) \leq K\}$. Now letting $K \to \infty$ (through integers) completes the proof. □

Theorem 8.5.2 (Conditional three-series theorem) *Let $(X_n, \mathcal{G}_n)$ be an adapted sequence and $(S_n = \sum_{i=1}^{n} X_i, \mathcal{G}_n)_{n \geq 1}$ be a martingale. Let $c > 0$ be any constant. Then S_n converges almost surely on the set where all of the following hold:*

(a) $\sum_{i=1}^{\infty} P\{|X_i| \geq c | \mathcal{G}_{i-1}\} < \infty$.

(b) $\sum_{i=1}^{\infty} E(X_i \mathbf{1}_{\{|X_i| \leq c\}} | \mathcal{G}_{i-1})$ *converges.*

(c) $\sum_{i=1}^{\infty} \left[E(X_i^2 \mathbf{1}_{\{|X_i| \leq c\}} | \mathcal{G}_{i-1}) - [E(X_i \mathbf{1}_{\{|X_i| \leq c\}} | \mathcal{G}_{i-1})]^2 \right]$ *converges.* ◆

Proof Define
$$Y_i := X_i \mathbf{1}_{\{|X_i| \leq c\}} - E(X_i \mathbf{1}_{\{|X_i| \leq c\}} | \mathcal{G}_{i-1}), \; i \geq 1.$$

Let A be the set where Conditions (a), (b) and (c) all hold. Using (a), Theorem 8.4.1, and (b), on the set A we have

$$\Big\{ \sum_{i=1}^{\infty} X_i \text{ converges} \Big\} = \Big\{ \sum_{i=1}^{\infty} X_i \mathbf{1}_{\{|X_i| \leq c\}} \text{ converges} \Big\} = \Big\{ \sum_{i=1}^{\infty} Y_i \text{ converges} \Big\}.$$

Observe that $(\sum_{i=1}^{n} Y_i, \mathcal{G}_n)_{n \geq 1}$ is a zero mean martingale, and

$$E(Y_i^2 | \mathcal{G}_{i-1}) = E(X_i^2 \mathbf{1}_{\{|X_i| \leq c\}} | \mathcal{G}_{i-1}) - [E(X_i \mathbf{1}_{\{|X_i| \leq c\}} | \mathcal{G}_{i-1})]^2.$$

Hence by Theorem 8.5.1, $\sum_{i=1}^{\infty} Y_i$ converges a.s. on A. This completes the proof. □

Exercise 8.5.1 Let $\{Z_n\}_{n \geq 1}$ be independent random variables, and $P(Z_n = \pm n^{-1/2}) = 1/2$. Let $Z_0 = 0$.
(a) Let $X_i = Z_i - Z_{i-1}$, $i \geq 1$. Show that $\sum_{i=1}^{\infty} X_i$ converges almost surely.
(b) Let $Y_i = Z_i + Z_{i-1}$, $i \geq 1$. Show that $\sum_{i=1}^{\infty} Y_i$ diverges almost surely.

Exercise 8.5.2 (*Kolmogorov's Three-Series Theorem*) Let $\{X_n\}$ be a sequence of independent random variables. Using the conditional three-series Theorem 8.5.2, show that $\sum_{n=1}^{\infty} X_n$ converges almost surely if and only for some $K > 0$, the following three properties hold:

(i) $\sum_{n=1}^{\infty} P(|X_n| > K) < \infty$,
(ii) $\sum_{n=1}^{\infty} E(X_n \mathbf{1}_{\{|X_n| < K\}}) < \infty$,
(iii) $\sum_{n=1}^{\infty} \text{Var}(X_n \mathbf{1}_{\{|X_n| < K\}}) < \infty$.

8.6 Toeplitz and Kronecker Lemma

These extremely useful lemmas shall be used in Sect. 8.7.

Theorem 8.6.1 (Toeplitz Lemma) *Suppose $\{k_n\}_{n \geq 1}$ is a sequence of integers and $\{a_{ni}\}_{n \geq 1, 1 \leq i \leq k_n}$ and $\{x_i\}_{i \geq 1}$ are real numbers such that*

8.6 Toeplitz and Kronecker Lemma

(a) $\lim_{n\to\infty} a_{ni} = 0$ for all $i \geq 1$.
(b) $\sup_n \sum_{i=1}^{k_n} |a_{ni}| \leq C < \infty$.
(c) $x_n \to 0$.

Then

(i) $\lim_{n\to\infty} \sum_{i=1}^{k_n} a_{ni} x_i = 0$.

(ii) *Suppose further that,* $\lim_{n\to\infty} \sum_{i=1}^{k_n} a_{ni} = 1$, *and* $\lim_{n\to\infty} x_n = x$. *Then* $\sum_{i=1}^{k_n} a_{ni} x_i \to x$.

(iii) *In particular, if* $x_n \to x$ *and* $\{a_i\}$ *are positive numbers such that* $b_n = \sum_{i=1}^{n} a_i \to \infty$, *then* $\sum_{i=1}^{n} a_i x_i / b_n \to x$. ♦

Proof (i) Fix $\varepsilon > 0$. Choose n_ε such that $|x_n| \leq \varepsilon$ for all $n \geq n_\varepsilon$. Then

$$|\sum_{i=1}^{k_n} a_{ni} x_i| \leq |\sum_{i=1}^{n_\varepsilon} a_{ni} x_i| + |\sum_{i > n_\varepsilon} a_{ni} x_i|$$

$$\leq |\sum_{i=1}^{n_\varepsilon} a_{ni} x_i| + C\varepsilon \text{ (by Condition (b))}.$$

By Condition (a), the first term on the right goes to 0. This completes the proof of (i) since ε was arbitrary. Part (ii) follows easily after writing

$$\sum_{i=1}^{k_n} a_{ni} x_i = x \sum_{i=1}^{k_n} a_{ni} + \sum_{i=1}^{k_n} a_{ni}(x_i - x).$$

(iii) This is immediate, once we choose $a_{ni} = a_i/b_n$ and use Part (ii). □

Theorem 8.6.2 (Kronecker's Lemma) *Let* $\{x_n\}_{n \geq 1}$ *be a sequence of real numbers such that* $\sum_{n=1}^{\infty} x_n$ *converges. Let* $\{b_n\}_{n \geq 1}$ *be a sequence of positive numbers increasing to* ∞. *Then*

$$\lim_{n\to\infty} \sum_{i=1}^{n} b_i x_i / b_n = 0.$$

♦

Proof Let $b_0 := 0$, $a_i := b_i - b_{i-1}, i \geq 1$, $s_{n+1} = \sum_{i=1}^{n} x_i \to s$ (say). Now,

$$\frac{\sum_{i=1}^{n} b_i x_i}{b_n} = \frac{1}{b_n} \sum_{i=1}^{n} b_i(s_{i+1} - s_i)$$

$$= s_{n+1} - \frac{1}{b_n} \sum_{i=1}^{n} a_i s_i \to 0 \text{ (by Theorem 8.6.1(iii))}.$$

□

8.7 Strong Law of Large Numbers for Martingales

There are many strong law type results for martingales. We borrow one from Hall and Heyde (1980) that implies the classical SLLN.

Theorem 8.7.1 (SLLN for martingales) *Let $(X_n, \mathcal{G}_n)_{n \geq 1}$ be an adapted sequence. Let X be a random variable such that $\mathrm{E}(|X|) < \infty$, and there exists a constant c such that*

$$P(|X_n| > x) \leq c P(|X| > x), \text{ for all } x > 0 \text{ and } n \geq 1. \tag{8.10}$$

(a) Then

$$\frac{1}{n} \sum_{i=1}^{n} (X_i - \mathrm{E}(X_i | \mathcal{G}_{i-1})) \to 0 \text{ in probability}. \tag{8.11}$$

(b) If $\mathrm{E}(|X| \log^+ |X|) < \infty$ or X_n's are independent, (8.11) holds a.s. ◆

Proof Let $Y_n := X_n \mathbf{1}_{\{|X_n| \leq n\}}$, $n \geq 1$. Observe that

$$\sum_{n=1}^{\infty} \frac{1}{n^2} \mathrm{E}\left((Y_n - \mathrm{E}(Y_n | \mathcal{G}_{n-1}))^2\right)$$

$$\leq \sum_{n=1}^{\infty} \frac{1}{n^2} \mathrm{E}(Y_n^2)$$

$$\leq 2 \sum_{n=1}^{\infty} \frac{1}{n^2} \int_0^n x P(|X_n| > x) dx$$

$$\leq 2c \sum_{n=1}^{\infty} \frac{1}{n^2} \int_0^n x P(|X| > x) dx$$

$$= 2c \sum_{n=1}^{\infty} \frac{1}{n^2} \sum_{i=1}^{n} \int_{\{i-1 \leq x \leq i\}} x P(|X| > x) dx$$

$$\leq 2c \sum_{n=1}^{\infty} \frac{1}{n^2} \sum_{i=1}^{n} i P(|X| > i-1)$$

$$\leq 2c \sum_{i=1}^{\infty} i P(|X| > i-1) \sum_{n=i}^{\infty} \frac{1}{n^2}$$

$$\leq 4c \sum_{i=1}^{\infty} P(|X| > i-1) < \infty \text{ (since } \mathrm{E}(|X|) < \infty).$$

This shows that

$$\sum_{n=1}^{\infty} \frac{1}{n^2} \mathrm{E}\left((Y_n - \mathrm{E}(Y_n | \mathcal{G}_{n-1}))^2\right) < \infty.$$

8.7 Strong Law of Large Numbers for Martingales

It follows from Theorem 8.5.1 that $\sum_n \frac{1}{n}(Y_n - E(Y_n|\mathcal{G}_{n-1}))$ is finite a.s. Hence using Kronecker's lemma 8.6.2,

$$\frac{1}{n}\sum_{i=1}^{n}(Y_i - E(Y_i|\mathcal{G}_{i-1})) \to 0 \text{ almost surely.} \quad (8.12)$$

Now

$$\sum_{n=1}^{\infty} P(X_n \neq Y_n) = \sum_{n=1}^{\infty} P(|X_n| > n) \leq c\sum_{n=1}^{\infty} P(|X| > n) < \infty.$$

Thus the sequences $\{X_n\}$ and $\{Y_n\}$ are *tail equivalent*. Hence using (8.12), it follows that

$$\frac{1}{n}\sum_{i=1}^{n}(X_i - E(Y_i|\mathcal{G}_{i-1})) \to 0 \text{ almost surely.} \quad (8.13)$$

Using the tower property of conditional expectation, we have

$$\frac{1}{n}E\left|\sum_{i=1}^{n}E(X_i\mathbf{1}_{\{|X_i|>i\}}|\mathcal{G}_{i-1})\right| \leq \frac{1}{n}\sum_{i=1}^{n}E(|X_i|\mathbf{1}_{\{|X_i|>i\}})$$

$$\leq \frac{1}{n}\sum_{i=1}^{n}\int_{i}^{\infty}P(|X_i| > x)dx$$

$$\leq \frac{c}{n}\sum_{i=1}^{n}\int_{i}^{\infty}P(|X| > x)dx \to 0.$$

The last convergence follows since $\int_{n}^{\infty}P(|X| > x)dx \to 0$.
This implies that

$$\frac{1}{n}\sum_{i=1}^{n}E(X_i - Y_i|\mathcal{G}_{i-1}) = \frac{1}{n}\sum_{i=1}^{n}E(X_i\mathbf{1}_{\{|X_i|>i\}}|\mathcal{G}_{i-1}) \to 0 \text{ in probability.} \quad (8.14)$$

This proves (8.11).
(b) When X_n are independent, $E(X_n - Y_n|\mathcal{G}_{n-1})$ is constant, and hence (8.14) holds a.s.

Suppose now that $E(|X|\log^+(|X|)) < \infty$. Then we first show that

$$\sum_{n=1}^{\infty}\frac{1}{n}E(|X_n|\mathbf{1}_{\{|X_n|>n\}}|\mathcal{G}_{n-1}) < \infty \text{ almost surely.} \quad (8.15)$$

$$\sum_{n=1}^{\infty} \frac{1}{n} E(|X_n| \mathbf{1}_{\{|X_n|>n\}}) = \sum_{n=1}^{\infty} \frac{1}{n} \int_n^{\infty} P(|X_n| > x) dx$$

$$\leq c \sum_{n=1}^{\infty} \frac{1}{n} \sum_{i=n}^{\infty} \int_{\{i \leq x \leq i+1\}} P(|X| > x) dx \text{ (using (8.10))}$$

$$\leq c \sum_{i=1}^{\infty} P(|X| > i) \sum_{n=1}^{i} \frac{1}{n}$$

$$\leq c \sum_{i=1}^{\infty} (1 + \log i) P(|X| > i) < \infty.$$

Hence (8.15) holds. Using Kronecker's Lemma (Theorem 8.6.2),

$$\frac{1}{n} \sum_{i=1}^{n} E(|X_i| \mathbf{1}_{\{|X_i|>i\}} | \mathcal{G}_{i-1}) \to 0 \text{ almost surely}.$$

This in turn implies

$$\frac{1}{n} \sum_{i=1}^{n} E(X_i - Y_i | \mathcal{G}_{i-1}) = \frac{1}{n} \sum_{i=1}^{n} E\left(X_i \mathbf{1}_{\{|X_i|>i|\mathcal{G}_{i-1}\}}\right) \to 0 \text{ almost surely.}$$

Hence using (8.13) the result follows. □

Corollary 8.7.1 *Let* $\{X_n\}_{n \geq 1}$ *be zero mean square integrable random variables. Let* $\mathcal{G}_{[r,s]} := \sigma(X_r, \ldots, X_s)$, $1 \leq r \leq s < \infty$. *Suppose there exists an integer* N *and a function* $f(\cdot)$ *of positive integers, decreasing to* 0, *such that for all* $n \geq N$ *and* $m \geq 1$,

$$\left| E(X_{m+n} | \mathcal{G}_{[1,m]}) \right| \leq f(n) E|X_{m+n}| \text{ almost surely.} \tag{8.16}$$

Suppose there exists a sequence of positive constants $b_n \uparrow \infty$ *such that*

$$\sum_{n=1}^{\infty} \frac{1}{b_n^2} E(X_n^2) < \infty \text{ and } \sup_n \frac{1}{b_n} \sum_{i=1}^{n} E|X_i| < \infty.$$

Then $b_n^{-1} \sum_{i=1}^{n} X_i \to 0$, *almost surely.* ♦

The above result gives a sufficient *mixing condition* for a strong law of large numbers for martingales.

Proof Since f is decreasing to 0, given $\varepsilon > 0$, choose $n_0 \geq N$ such that $f(n_0) < \varepsilon$. From (8.16) we have for $i, j \geq 1$,

$$\left|E(X_{in_0+j}|X_{n_0+j}, X_{2n_0+j}, \ldots, X_{(i-1)n_0+j})\right|$$
$$\leq \left|E[E(X_{in_0+j}|X_1, X_2, \ldots, X_{(i-1)n_0+j})|X_{n_0+j}, X_{2n_0+j}, \ldots, X_{(i-1)n_0+j})]\right|$$
$$\leq f(n_0)E|X_{(i-1)n_0+j}|.$$

If $n \geq n_0$, choose a non-negative q and r such that $0 \leq r \leq n_0 - 1$, and $n = qn_0 + r$. Then

$$\frac{1}{b_n}\sum_{i=1}^{n} X_i = \frac{1}{b_n}\sum_{i=1}^{n_0} X_i + \frac{1}{b_n}\sum_{i=1}^{q-1}\sum_{j=1}^{n_0-1} X_{in_0+j} + \frac{1}{b_n}\sum_{j=1}^{r} X_{qn_0+j}. \tag{8.17}$$

The first term converges to 0 almost surely as it involves a sum of finitely many elements. The sum of the last two terms is bounded by

$$\sum_{j=1}^{n_0-1} \frac{1}{b_n}\sum_{i=1}^{q-1} \left|E(X_{in_0+j}) - E(X_{in_0+j}|X_{n_0+j}, \ldots, X_{(i-1)n_0+j}))\right|$$

$$+ \sum_{j=1}^{r} \frac{1}{b_n} \left|X_{qn_0+j} - E\left(X_{qn_0+j}|X_{n_0+j}, \ldots, X_{(q-1)n_0+j}\right)\right|$$

$$+ \frac{f(n_0)}{b_n} \sum_{i=n_0+1}^{n} E|X_i|.$$

Since $\sum_{n=1}^{\infty} b_n^{-2}E(X_n^2) < \infty$, Theorems 8.5.1 and 8.6.2 imply that the first term converges to 0. The second term goes to 0 as r is fixed, and the last term is dominated by $\sup_{n\geq 1} b_n^{-1}\sum_{i=1}^{n} E|X_i| < \infty$. Hence,

$$\limsup_{n\to\infty} |\frac{1}{b_n}\sum_{i=1}^{n} X_i| \leq \varepsilon(\sup_{n\geq 1} \frac{1}{b_n}\sum_{i=1}^{n} E|X_i|) \text{ almost surely.}$$

Hence the result follows. $\square$

8.8 Burkholder–Davis–Gundy Inequalities

The famous inequality of Burkholder and Gundy (1970) provides bounds for the L^p norm of the maximum of (discrete time) martingales for $p > 1$. In the same year, Davis (1970) established a version of the inequality for $p = 1$. The two inequalities are stated below in Theorem 8.8.1. We shall prove the inequality for $p > 1$. The proof for $p = 1$ is significantly more difficult, and we refer the readers to Chow and Teicher (2003).

For various other results around these inequalities for discrete time martingales, we refer to Chow and Teicher (2003). Extension of the inequalities to continuous-time martingales was accomplished by Lenglart (1983). Since we do not cover continuous time martingales in this book, we direct our readers to Revuz and Yor (2013) for further information. These inequalities are now collectively referred to as Burkholder–Davis–Gundy inequalities.

Theorem 8.8.1 (Burkholder–Davis–Gundy) *(i) For $p \in (1, \infty)$, there are constants $c_p, C_p \in (0, \infty)$ such that for any martingale $(X_n, \mathcal{G}_n)$ with $X_0 = 0$, we have for $n \geq 1$,*

$$c_p \mathrm{E}\Big(\big(\sum_{i=1}^{n}(X_i - X_{i-1})^2\big)^{p/2}\Big) \leq \mathrm{E}\big(|X_n|^p\big)$$

$$\leq \mathrm{E}\big(\max_{1 \leq i \leq n} |X_i|^p\big)$$

$$\leq C_p \mathrm{E}\Big(\big(\sum_{i=1}^{n}(X_i - X_{i-1})^2\big)^{p/2}\Big).$$

(ii) There exists constants $c_1, C_1 \in (0, \infty)$ such that for any L^1 bounded martingale $(X_n, \mathcal{G}_n)$ with $X_0 = 0$, we have for $n \geq 1$,

$$c_1 \mathrm{E}\Big(\big(\sum_{i=1}^{n}(X_i - X_{i-1})^2\big)^{1/2}\Big) \leq \mathrm{E}\big(\sup_{1 \leq i \leq n} |X_i|\big)$$

$$\leq C_1 \mathrm{E}\Big(\big(\sum_{i=1}^{n}(X_i - X_{i-1})^2\big)^{1/2}\Big).$$

◆

Remark 8.8.1 It may observed that when $p = 2$ and $\mathrm{E}(X_n^2) < \infty$, the leftmost inequality in Part (i) is an equality with $c_2 = 1$. If $\{A_n\}$ is the quadratic variation process of $\{X_n\}$ as in Exercise 4.7.5, that is,

$$A_n = \sum_{i=2}^{n} \mathrm{E}((X_i - X_{i-1})^2 | \mathcal{G}_{i-1}). \tag{8.18}$$

Then Theorem 8.8.1(i) for $p = 2$ can be strengthened to

$$\mathrm{E}(X_1^2 + A_n) = \mathrm{E}(X_n^2) \leq \mathrm{E}(\max_{1 \leq i \leq n} X_i^2) \leq C_2 \mathrm{E}(X_1^2 + A_n). \tag{8.19}$$

●

The proof of Theorem 8.8.1 relies on the following two lemmas.

8.8 Burkholder–Davis–Gundy Inequalities

Lemma 8.8.1 *Let $(X_n, \mathcal{G}_n)_{n\geq 0}$ be a non-negative sub-martingale with $X_0 = 0$. Then,*

$$P\Big(\sum_{i=1}^{n-1}(X_i - X_{i-1})^2 > \lambda^2, \max_{1\leq i\leq n} X_i \leq \lambda\Big) \leq \frac{2}{\lambda}E(X_n), \text{ for } \lambda > 0, \ n \geq 1.$$

♦

Proof Using $X_0 = 0$ write

$$X_n^2 = \Big(\sum_{i=1}^{n}(X_i - X_{i-1})\Big)^2$$

$$= \sum_{i=1}^{n}(X_i - X_{i-1})^2 + 2\sum_{1\leq i<j\leq n}(X_i - X_{i-1})(X_j - X_{j-1}).$$

Thus,

$$\frac{1}{2}\Big(X_n^2 + \sum_{i=1}^{n}(X_i - X_{i-1})^2\Big) = \sum_{1\leq i\leq j\leq n}(X_i - X_{i-1})(X_j - X_{j-1})$$

$$= \sum_{i=1}^{n}(X_i - X_{i-1})(X_n - X_{i-1})$$

$$= X_n^2 - \sum_{i=1}^{n}X_{i-1}(X_i - X_{i-1})$$

$$= X_n X_{n+1} - \sum_{i=1}^{n+1}X_{i-1}(X_i - X_{i-1}).$$

This immediately implies that for all $n \geq 1$,

$$\frac{1}{2}\sum_{i=1}^{n}(X_i - X_{i-1})^2 \leq X_n X_{n+1} - \sum_{i=1}^{n+1}X_{i-1}(X_i - X_{i-1}). \quad (8.20)$$

Let $T = \inf\{k \geq 1 : X_k > \lambda\}$ and $\tau = T \wedge n$, $n \geq 1$. Note that $\tau \leq n$. Replacing n by $\tau - 1$ in (8.20),

$$\frac{1}{2}\sum_{i=1}^{\tau-1}(X_i - X_{i-1})^2 \leq X_{\tau-1}X_\tau - \sum_{i=1}^{n}\mathbf{1}_{\{\tau\geq i\}}X_{i-1}(X_i - X_{i-1}). \quad (8.21)$$

Since for $i = 1, \ldots, n$,

$$\mathbf{1}_{\{\tau\geq i\}}|X_{i-1}(X_i - X_{i-1})| \leq \lambda(X_i + \lambda),$$

$\mathbf{1}_{\{\tau \geq i\}} X_{i-1}(X_i - X_{i-1})$ is integrable. Furthermore, τ is a stopping time and consequently $\{\tau \geq i\} \in \mathcal{G}_{i-1}$. Now since $X_{i-1} \geq 0$ and $(X_n, \mathcal{G}_n)$ is a sub-martingale, we have almost surely,

$$\mathrm{E}\big(\mathbf{1}_{\{\tau \geq i\}} X_{i-1}(X_i - X_{i-1}) | \mathcal{G}_{i-1}\big) = \mathbf{1}_{\{\tau \geq i\}} X_{i-1} \mathrm{E}\big(X_i - X_{i-1} | \mathcal{G}_{i-1}\big) \geq 0.$$

Taking expectation and substituting this in (8.21) yields

$$\begin{aligned}
\mathrm{E}\Big(\sum_{i=1}^{\tau-1}(X_i - X_{i-1})^2\Big) &\leq 2\mathrm{E}(X_\tau X_{\tau-1}) \\
&\leq 2\lambda \mathrm{E}(X_\tau) \\
&\leq 2\lambda \mathrm{E}(X_n) \quad \text{(by Theorem 4.3.2)}. \quad (8.22)
\end{aligned}$$

Clearly $\tau = n$ if $\max_{1 \leq i \leq n} |X_i| \leq \lambda$. Thus,

$$P\Big(\sum_{i=1}^{n-1}(X_i - X_{i-1})^2 > \lambda^2, \max_{1 \leq i \leq n} |X_i| \leq \lambda\Big) \leq P\Big(\sum_{i=1}^{\tau-1}(X_i - X_{i-1})^2 > \lambda^2\Big)$$

$$\leq \lambda^{-2} \mathrm{E}\Big(\sum_{i=1}^{\tau-1}(X_i - X_{i-1})^2\Big).$$

Using (8.22), the proof is complete. $\square$

Lemma 8.8.2 *For $1 < p < \infty$ there exists $\tilde{A}_p \in (0, \infty)$ such that for any non-negative sub-martingale $(X_n, \mathcal{G}_n)_{n \geq 1}$ with $X_0 = 0$,*

$$\mathrm{E}\Big((\sum_{i=1}^n (X_i - X_{i-1})^2)^{p/2}\Big) \leq \tilde{A}_p \mathrm{E}(X_n^p), \; n \geq 1.$$

♦

Proof Fix $\lambda \in (0, \infty)$ and $\beta \in (1, \infty)$ and let

$$Y_n = X_n \mathbf{1}_{\{\sum_{j=1}^n (X_j - X_{j-1})^2 > \lambda^2\}}, \; n \geq 0.$$

For $n \geq 1$,

$$\begin{aligned}
\mathrm{E}(Y_{n+1}|\mathcal{G}_n) &\geq \mathrm{E}\big(X_{n+1} \mathbf{1}_{\{\sum_{j=1}^n (X_j - X_{j-1})^2 > \lambda^2\}} | \mathcal{G}_n\big) \\
&= \mathbf{1}_{\{\sum_{j=1}^n (X_j - X_{j-1})^2 > \lambda^2\}} \mathrm{E}(X_{n+1}|\mathcal{G}_n) \\
&\geq Y_n.
\end{aligned}$$

Thus, $(Y_n, \mathcal{G}_n)_{n \geq 0}$ is also a non-negative sub-martingale with $Y_0 = 0$. Lemma 8.8.1 implies

8.8 Burkholder–Davis–Gundy Inequalities

$$P\Big(\sum_{i=1}^{n-1}(Y_i - Y_{i-1})^2 > \lambda^2, \max_{1\le i \le n} Y_i \le \lambda\Big) \le \frac{2}{\lambda} E(Y_n), \; n \ge 1. \tag{8.23}$$

Let

$$T = \inf\{n \ge 1 : \sum_{j=1}^{n}(X_j - X_{j-1})^2 > \lambda^2\}.$$

As $\beta > 1$,

$$\{\sum_{j=1}^{n}(X_j - X_{j-1})^2 > \lambda^2 \beta^2\} \subset \{T \le n\}. \tag{8.24}$$

Further, $Y_i \le X_i$ for all i. Thus, on the set

$$\{\sum_{j=1}^{n}(X_j - X_{j-1})^2 > \lambda^2 \beta^2, \max_{1\le i \le n} X_i \le \lambda\} \subset \{T \le n, \max_{1\le i \le n} X_i \le \lambda\}, \tag{8.25}$$

we have

$$\max_{1\le i \le n} Y_i \le \lambda, \text{ and}$$

$$\begin{aligned}
|X_T - X_{T-1}| &\le X_T \vee X_{T-1} \\
&\le \max_{1\le i \le n} X_i \\
&\le \lambda.
\end{aligned}$$

Besides, on the set in left side of (8.25),

$$\begin{aligned}
\beta^2 \lambda^2 &< \sum_{j=1}^{n}(X_j - X_{j-1})^2 \\
&= \sum_{j=1}^{T-1}(X_j - X_{j-1})^2 + (X_T - X_{T-1})^2 + \sum_{j=T+1}^{n}(X_j - X_{j-1})^2 \\
&\le 2\lambda^2 + \sum_{j=T+1}^{n}(X_j - X_{j-1})^2 \text{ (using definition of } T \text{ and (8.24))} \\
&\le 2\lambda^2 + \sum_{j=1}^{n}(Y_j - Y_{j-1})^2 \text{ (since } Y_j = X_j \text{ for } j \ge T+1).
\end{aligned}$$

Taking $\beta = \sqrt{3}$, we thus get

$$\{\sum_{j=1}^{n}(X_j - X_{j-1})^2 > 3\lambda^2, \max_{1\leq i\leq n} X_i \leq \lambda\}$$
$$\subset \{\sum_{j=1}^{n}(Y_j - Y_{j-1})^2 > \lambda^2, \max_{1\leq i\leq n} Y_i \leq \lambda\}.$$

We can combine this with (8.23) to conclude

$$P\Big(\sum_{j=1}^{n}(X_j - X_{j-1})^2 > 3\lambda^2, \max_{1\leq i\leq n} X_i \leq \lambda\Big) \leq \frac{2}{\lambda}E(Y_n). \qquad (8.26)$$

Recall Theorem 4.5.2 which implies

$$P\Big(\max_{1\leq i\leq n} X_i > \lambda\Big) \leq \frac{1}{\lambda}E\big(X_n \mathbf{1}_{\{\max_{1\leq i\leq n} X_i \geq \lambda\}}\big).$$

Let
$$Z_n = \max\{X_1, \ldots, X_n, (\sum_{j=1}^{n}(X_j - X_{j-1})^2)^{1/2}\}, n \geq 1. \qquad (8.27)$$

Then the above in conjunction with (8.26) implies

$$P\big(\{\sum_{j=1}^{n}(X_j - X_{j-1})^2 > 3\lambda^2\} \cup \{\max_{1\leq i\leq n} X_i > \lambda\}\big) \qquad (8.28)$$
$$\leq \frac{2}{\lambda}E(Y_n) + \frac{1}{\lambda}E\big(X_n \mathbf{1}_{\{\max_{1\leq i\leq n} X_i \geq \lambda\}}\big)$$
$$= \frac{1}{\lambda}\big(2E\big(X_n \mathbf{1}_{\{\sum_{j=1}^{n}(X_j - X_{j-1})^2 > \lambda^2\}}\big) + E\big(X_n \mathbf{1}_{\{\max_{1\leq i\leq n} X_i \geq \lambda\}}\big)\big)$$
$$\leq \frac{3}{\lambda}E\big(X_n \mathbf{1}_{\{Z_n \geq \lambda\}}\big). \qquad (8.29)$$

Note that

$$\frac{E(Z_n^p)}{3^{p/2}}$$
$$= \int_0^\infty P(Z_n > \sqrt{3}s^{1/p})\,ds = p\int_0^\infty P(Z_n > \sqrt{3}\lambda)\lambda^{p-1}\,d\lambda$$
$$= p\int_0^\infty P\big(\{\sum_{j=1}^{n}(X_j - X_{j-1})^2 > 3\lambda^2\} \cup \{\max_{1\leq i\leq n} X_i > \sqrt{3}\lambda\}\big)\lambda^{p-1}\,d\lambda$$

8.8 Burkholder–Davis–Gundy Inequalities

$$\le p \int_0^\infty P(\{\sum_{j=1}^n (X_j - X_{j-1})^2 > 3\lambda^2\} \cup \{\max_{1 \le i \le n} X_i > \lambda\}) \lambda^{p-1} \, d\lambda$$

$$\le 3p \int_0^\infty \mathrm{E}\left(X_n \mathbf{1}_{\{Z_n \ge \lambda\}}\right) \lambda^{p-2} \, d\lambda \quad \text{(by (8.28) and (8.29))}$$

$$\le 3p \int_0^\infty \int \mathbf{1}_{\{Z_n \ge \lambda\}} X_n \lambda^{p-2} \, dP \, d\lambda \quad \text{(by Fubini's theorem)}$$

$$= 3p \int X_n \int_0^{Z_n} \lambda^{p-2} \, d\lambda \, dP$$

$$= \frac{3p}{p-1} \mathrm{E}\left(X_n Z_n^{p-1}\right)$$

$$\le \frac{3p}{p-1} \left(\mathrm{E}(X_n^p)\right)^{1/p} \left(\mathrm{E}(Z_n^p)\right)^{(p-1)/p} \quad \text{(by Hölder's inequality)}.$$

By (8.27) and adjusting the terms using the last inequality, one can conclude,

$$3^{1+p/2} \frac{p}{p-1} \left(\mathrm{E}(X_n^p)\right)^{1/p} \ge \left(\mathrm{E}(Z_n^p)\right)^{1/p} \ge \left(\mathrm{E}\left(\left(\sum_{j=1}^n (X_j - X_{j-1})^2\right)^{p/2}\right)\right)^{1/p}.$$

Letting

$$\tilde{A}_p = \left(3^{1+p/2} \frac{p}{p-1}\right)^p,$$

the proof is complete. □

Proof of Theorem 8.8.1(i) In view of (5.7) obtained in the proof of Theorem 5.3.1, it suffices to show that there exist universal constants $c_p, C_p' \in (0, \infty)$ such that

$$c_p \mathrm{E}\left(\left(\sum_{i=1}^n (X_i - X_{i-1})^2\right)^{p/2}\right) \le \mathrm{E}(|X_n|^p)$$

$$\le C_p' \mathrm{E}\left(\left(\sum_{i=1}^n (X_i - X_{i-1})^2\right)^{p/2}\right). \quad (8.30)$$

We shall show that the left inequality of (8.30) holds with $c_p = (2^p \tilde{A}_p)^{-1}$ where $\tilde{A}_p$ is as in Lemma 8.8.2.

Fix n and let $Y_0 = Z_0 = 0$, and for $i = 1, \ldots, n$,

$$Y_i = \mathrm{E}(X_n^+ | \mathcal{G}_i) \text{ and}$$
$$Z_i = \mathrm{E}(X_n^- | \mathcal{G}_i).$$

Thus, $Y_i - Z_i = X_i$, $i = 0, \ldots, n$. Hence,

$$\Big(\sum_{i=1}^{n}(X_i - X_{i-1})^2\Big)^{1/2} \le \Big(\sum_{i=1}^{n}(Y_i - Y_{i-1})^2\Big)^{1/2} + \Big(\sum_{i=1}^{n}(Z_i - Z_{i-1})^2\Big)^{1/2}.$$

Denoting by $\|\cdot\|_p$ the L^p-norm, the non-negativity of the left side implies

$$\Big\|\Big(\sum_{i=1}^{n}(X_i - X_{i-1})^2\Big)^{1/2}\Big\|_p$$
$$\le \Big\|\Big(\sum_{i=1}^{n}(Y_i - Y_{i-1})^2\Big)^{1/2} + \Big(\sum_{i=1}^{n}(Z_i - Z_{i-1})^2\Big)^{1/2}\Big\|_p$$
$$\le \Big\|\Big(\sum_{i=1}^{n}(Y_i - Y_{i-1})^2\Big)^{1/2}\Big\|_p$$
$$+ \Big\|\Big(\sum_{i=1}^{n}(Z_i - Z_{i-1})^2\Big)^{1/2}\Big\|_p \quad \text{(by Minkowski's inequality)}.$$

Since $(Y_i, \mathcal{G}_i)_{0 \le i \le n}$ is a non-negative martingale and $Y_0 = 0$, Lemma 8.8.2 implies

$$\Big\|\Big(\sum_{i=1}^{n}(Y_i - Y_{i-1})^2\Big)^{1/2}\Big\|_p \le \tilde{A}_p^{1/p} \|Y_n\|_p \le \tilde{A}_p^{1/p} \|X_n\|_p,$$

and likewise for $(Z_i, \mathcal{G}_i)_{0 \le i \le n}$. The left inequality of (8.30) thus holds with $c_p = (2^p \tilde{A}_p)^{-1}$.

For the right inequality, assume without loss that $P(X_n \ne 0) > 0$ and

$$E\Big(\Big(\sum_{i=1}^{n}(X_i - X_{i-1})^2\Big)^{p/2}\Big) < \infty.$$

An implication of the above is that for $i = 1, \ldots, n$, $X_i - X_{i-1} \in L^p$. Since $X_0 = 0$, it follows inductively that $X_1, \ldots, X_n \in L^p$.

Let
$$W_n = \text{sgn}(X_n) |X_n|^{p-1} \|X_n\|_p^{1-p}.$$

Let q satisfy $\frac{1}{p} + \frac{1}{q} = 1$. Note that $\|W_n\|_q = 1$. Let $W_j = E(W_n | \mathcal{G}_j)$, $0 \le j < n$. Then $W_0, \ldots, W_{n-1} \in L^q$. Further, $X_n W_n = \|X_n\|_p^{1-p} |X_n|^p$, a consequence of which is $E(X_n W_n) = \|X_n\|_p$. Write

$$X_n W_n = X_{n-1} W_{n-1} + (X_n - X_{n-1}) W_{n-1}$$
$$+ X_{n-1}(W_n - W_{n-1}) + (X_n - X_{n-1})(W_n - W_{n-1}).$$

A standard argument via conditioning on $\mathcal{G}_{n-1}$ shows

8.8 Burkholder–Davis–Gundy Inequalities

$$E((X_n - X_{n-1})W_{n-1}) = 0 = E(X_{n-1}(W_n - W_{n-1})).$$

Inductively it follows that

$$E(X_n W_n) = E\Big(\sum_{i=1}^n (X_i - X_{i-1})(W_i - W_{i-1})\Big)$$

$$\leq E\Big(\Big(\sum_{i=1}^n (X_i - X_{i-1})^2\Big)^{1/2} \Big(\sum_{i=1}^n (W_i - W_{i-1})^2\Big)^{1/2}\Big)$$

$$\leq \Big\|\Big(\sum_{i=1}^n (X_i - X_{i-1})^2\Big)^{1/2}\Big\|_p \Big\|\Big(\sum_{i=1}^n (W_i - W_{i-1})^2\Big)^{1/2}\Big\|_q.$$

Putting everything together we get

$$\|X_n\|_p \leq \Big\|\Big(\sum_{i=1}^n (X_i - X_{i-1})^2\Big)^{1/2}\Big\|_p \Big\|\Big(\sum_{i=1}^n (W_i - W_{i-1})^2\Big)^{1/2}\Big\|_q$$

$$\leq c_q^{-1/q} \|W_n\|_q \Big\|\Big(\sum_{i=1}^n (X_i - X_{i-1})^2\Big)^{1/2}\Big\|_p,$$

the last line following from the left inequality of (8.30) which has already been proved.

Recalling that $\|W_n\|_q = 1$, the right inequality of (8.30) follows with $C'_p = c_q^{-p/q}$. This completes the proof. □

The next result is similar in spirit to Theorem 8.8.1. It compares a square integrable martingale with its quadratic variation asymptotically.

Theorem 8.8.2 *Let $(X_n, \mathcal{G}_n)_{n \geq 1}$ be a martingale where $E(X_n^2) < \infty$ for all $n \geq 1$. Let A_n be the quadratic variation of $\{X_n\}$, and A_∞ its limit. That is,*

$$A_n = \sum_{i=2}^n E((X_i - X_{i-1})^2 | \mathcal{G}_{i-1}), n \geq 1, \quad A_\infty = \lim_{n \to \infty} A_n. \tag{8.31}$$

(a) On the set $\{A_\infty < \infty\}$, $\lim_{n \to \infty} X_n$ exists a.s.

(b) On the set $\{A_\infty = \infty\}$, for every function $f : \mathbb{R}^+ \to \mathbb{R}^+$ which is non-decreasing and satisfies $\int_0^\infty (1 + f(t))^{-2} dt < \infty$,

$$\lim_{n \to \infty} \frac{1}{f(A_n)} X_n = 0 \text{ a.s.} \tag{8.32}$$

♦

Proof (a) Define for all $a > 0$,

$$\tau_a = \inf\{n \geq 1 : A_{n+1} > a\}.$$

Since A_{n+1} is $\mathcal{G}_n$ measurable for all n, τ_a is a stopping time. Thus $(Y_n, \mathcal{G}_n)$ is a martingale where $Y_n = X_{\tau_a \wedge n}$ for $n \geq 1$.

Clearly $A_{\tau_a} \leq a$. On $\{\tau_a = \infty\}$, this is immediate. On $\{\tau_a < \infty\}$, it follows from the definition. Further, $Y_n = Y_{n+1}$ for $n \geq \tau_a$.

Hence

$$E\left(\sum_{n=1}^{\infty}(Y_{n+1} - Y_n)^2\right)$$

$$= E\left(\sum_{n=1}^{\tau_a-1}(Y_{n+1} - Y_n)^2\right)$$

$$= E\left(\sum_{n=1}^{\tau_a-1}(X_{n+1} - X_n)^2\right)$$

$$= E\left(\sum_{i=2}^{\infty}(X_i - X_{i-1})^2 \mathbf{1}(\tau_a \geq i)\right)$$

$$= E\left(\sum_{i=2}^{\infty} E\left((X_i - X_{i-1})^2 | \mathcal{G}_{i-1}\right) \mathbf{1}(\tau_a \geq i)\right) \text{ (as } \{\tau_a \geq i\} \in \mathcal{G}_{i-1})$$

$$= E(A_{\tau_a}) \leq a.$$

Exercise 5.5.6 shows that Y_n converges a.s.

Since $X_n = Y_n$ on the set $\{\tau_a < \infty\}$ which is the same as $\{A_\infty \leq a\}$, it thus follows that

$$\lim_{n \to \infty} X_n \text{ exists a.s. on the set } \{A_\infty \leq a\}.$$

Letting $a \to \infty$ (through integers) completes the proof of Part (a).

(b) Let $f : \mathbb{R}^+ \to \mathbb{R}^+$ satisfy the hypotheses. Define

$$Z_n = \sum_{i=1}^{n-1} \frac{X_{i+1} - X_i}{1 + f(A_{i+1})}, n \geq 1.$$

Now $Z_{n+1} - Z_n = (X_{n+1} - X_n)/(1 + f(A_{n+1}))$ and A_{n+1} is $\mathcal{G}_n$-measurable. Hence $(Z_n, \mathcal{G}_n)_{n \geq 1}$ is a martingale. Moreover, for $n \geq 1$,

$$E\left((Z_{n+1} - Z_n)^2 | \mathcal{G}_n\right) = \frac{1}{(1 + f(A_{n+1}))^2} E\left((X_{n+1} - X_n)^2 | \mathcal{G}_n\right)$$

$$= \frac{A_{n+1} - A_n}{(1 + f(A_{n+1}))^2},$$

where the right equality follows from the definition of A_n given in (8.31). Taking expectation and summing over $n = 1, 2, \ldots$, we get

$$\sum_{n=1}^{\infty} E((Z_{n+1} - Z_n)^2) = E\left(\sum_{n=1}^{\infty} \frac{A_{n+1} - A_n}{(1 + f(A_{n+1}))^2}\right)$$

$$\leq E\left(\sum_{n=1}^{\infty} \int_{A_n}^{A_{n+1}} (1 + f(t))^{-2} dt\right) \text{ (}f \text{ non-decreasing)}$$

$$\leq \int_0^{\infty} (1 + f(t))^{-2} dt < \infty.$$

Exercise 5.5.6 shows Z_n converges a.s. to a finite limit. Thus, as $i \to \infty$,

$$\frac{X_{i+1} - X_i}{1 + f(A_{i+1})} \to 0 \text{ a.s.}$$

On the set $\{A_\infty = \infty\}$, the assumptions on f imply $f(A_n) \uparrow \infty$. Letting $x_i = \frac{X_{i+1} - X_i}{1 + f(A_{i+1})}$ and $b_n = 1 + f(A_{n+1})$, Theorem 8.6.2 shows that

$$b_n^{-1} \sum_{i=1}^{n} b_i x_i \to 0 \text{ on the set } \{A_\infty = \infty\}.$$

The left side equals $(X_{n+1} - X_1)/(1 + f(A_{n+1}))$. The proof of Part (b) now follows from the fact that $f(A_n) \uparrow \infty$ on the set $\{A_\infty = \infty\}$. □

Remark 8.8.2 Taking f to be the identity function in Theorem 8.8.2 shows that on the set $\{A_\infty = \infty\}$, $A_n^{-1} X_n \to 0$ a.s. ●

8.9 Branching Process: Kesten–Stigum Theorem

We investigate further the BGW process $\mathbf{Z} = (Z_i : i \geq 0)$ of Example 4.1.5. So, let ξ be an $\mathbb{N}_0 = 0, 1, 2, \ldots$-valued random variable. Let $(\xi_{n,i} : i \geq 1)_{n \geq 0}$ be iid each having the same distribution as that of ξ.
Then

$$Z_0 = 1 \text{ and } Z_n = 1_{\{Z_{n-1} \geq 1\}} \sum_{i=1}^{Z_{n-1}} \xi_{n-1,i} \text{ for all } n \geq 1. \tag{8.33}$$

The size Z_i of each generation depends on the previous generation, and if $Z_i = 0$ for some i, then the subsequent generations are also of size 0.

It can be easily checked that $P(\lim_{n\to\infty} Z_n \in \{0, \infty\}) = 1$ that is, Z_n either converges to zero or diverges to ∞. A natural question is whether we can compute $P(\lim_{n\to\infty} Z_n = 0)$.

Observe that $\lim_{n\to\infty} Z_n = 0$ if and only if $Z_n = 0$ eventually, that is, $\{\lim_{n\to\infty} Z_n = 0\} = \cup_{n\geq 1}\{Z_n = 0\}$. This event is referred to as the *extinction event* and the complementary event $\cap_{n\geq 0}\{Z_n \geq 1\}$ is called the *survival event*.

Let us assume that $m = E(\xi) < \infty$. Then it is easy to check that $E(Z_n|Z_{n-1}) = mZ_{n-1}$ almost surely for all $n \geq 1$ that is, $E(Z_n) = m^n$ for all $n \geq 0$.

If $m < 1$, then by Fatou's lemma,

$$E(\liminf_{n\to\infty} Z_n) \leq \liminf_{n\to\infty} E(Z_n) = 0.$$

So, it follows that $P(\lim_{n\to\infty} Z_n = 0) = 1$ if $m < 1$.

If $m = 1$, Fatou's lemma implies $E(\liminf_{n\to\infty} Z_n) \leq 1$, and hence the survival probability must be zero.

If $m > 1$, we need a more detailed study, involving the probability generating function of Z_n to conclude that the survival probability is positive. See Athreya and Ney (2004) and Asmussen and Hering (1983) for a detailed discussion on this.

Define

$$W_n := m^{-n} Z_n, \ n \geq 0.$$

Lemma 8.9.1 *Let $(\mathcal{G}_n : n \geq 1)$ denote the natural filtration of $\mathbf{Z}$. Then $((W_n, \mathcal{G}_n) : n \geq 0)$ is a non-negative martingale, and as a consequence, $W = \lim_{n\to\infty} W_n$ exists almost surely.* ♦

The above lemma is a direct consequence of the martingale convergence Theorem 5.2.1. But it does not imply that $P(W > 0) > 0$.

Exercise 8.9.1 For the BGW process, suppose that $E(\xi^2) < \infty$. Then show that

$$E(Z_n^2) = m^{2n}\left(1 + \text{Var}(\xi)\sum_{i=2}^{n+1} m^{-i}\right). \tag{8.34}$$

Using this, show that $P(W > 0) > 0$ and $E(W) = 1$.

Kesten and Stigum (1966) showed that the second moment condition in Exercise 8.9.1 can be reduced. Their proof relies on a truncation technique where the martingale $(W_n = m^{-n} Z_n : n \geq 1)$ is approximated by an L^2-martingale. Asmussen and Hering (1983) gave four different proofs. We present the one which uses Kolmogorov's three-series theorem (see Exercise 8.5.2).

Theorem 8.9.1 (Kesten–Stigum Theorem) *Suppose $Z_0 = 1$ almost surely. Then the following assertions are equivalent:*

(1) $E(\xi \log_+ \xi) < \infty$ where $\log_+ x = \log(x \vee 1)$.
(2) $P(W > 0) = P(\cap_{n\geq 1}\{Z_n \geq 1\}) > 0$.

8.9 Branching Process: Kesten–Stigum Theorem

(3) $E(W) = 1$.
(4) $\lim_{n\to\infty} E(|W_n - W|) = 0$.
(5) *The sequence* $(W_n : n \geq 0)$ *is uniformly integrable*.
(6) $E(\sup_{n\geq 1} W_n) < \infty$. ♦

Proof The equivalence of Conditions (3), (4) and (5) is immediate. It is obvious that (3) implies (2), and (6) implies (5). To complete the proof we show that: (2) implies (6), (1) implies (3), and (2) implies (1).

Step 1: Condition (2) $\Rightarrow$ Condition (6). We first derive a simple consequence of the branching property. Fix an integer N such that $Z_N > 0$. Consider the sub-trees rooted at the Nth generation. Let $Z_{N+k}^{(N,i)}$ denote the number of descendants in the $(N+k)$th generation of the ith particle in the Nth generation. Then conditioned on Z_N, $(Z_{N+k}^{(N,i)} : 1 \leq i \leq Z_N)$ is a collection of independent copies of Z_k. We further define
$$W^{(N,i)} := \lim_{k\to\infty} m^{-k} Z_{k+N}^{(N,i)} \text{ almost surely.}$$

Conditioned on Z_N, $(W^{(N,i)} : 1 \leq i \leq Z_N)$ are iid copies of W.

To show that $E(\sup_{n\geq 1} W_n) < \infty$, we derive an upper bound for the tail probability of $\sup_{n\geq 1} W_n$ in terms of the tail probability of W.

Observe that for any $x > 0$.

$$\{\sup_{n\geq 1} W_n > x\} = \bigcup_{n\geq 1}\left(\{W_n > x\} \cap \bigcap_{j=1}^{n-1}\{W_j \leq x\}\right) =: \bigcup_{n=1}^{\infty} \mathsf{F}_n(x). \tag{8.35}$$

Condition (2) and Fatou's lemma imply $E(W) \in (0, 1]$. Let $(\hat{W}_i : i \geq 1)$ be iid copies of W. Fix $g \in (0, \mathbb{E}(W))$. Then SLLN can be used to conclude that there exists $p \in (0, 1)$ such that

$$P\left(n^{-1}\sum_{i=1}^n \hat{W}_i \geq g\right) \geq p \text{ for all } n \geq 1. \tag{8.36}$$

We now use the branching property to observe that

$P(W > gx \mid \mathsf{F}_n(x))$

$= P\left(m^{-n} \lim_{k\to\infty} \sum_{i=1}^{Z_n}(m^{-k} Z_{n+k}^{(n,i)}) > gx \mid \{Z_n > \lfloor m^n x\rfloor + 1\} \cap_{j=1}^{n-1}\{W_j < x\}\right)$

$\geq P\left(m^{-n}\sum_{i=1}^{\lfloor m^n x\rfloor+1} W^{(n,i)} > gx \mid \mathsf{F}_n(x)\right)$

$\geq P\left(\sum_{i=1}^{\lfloor m^n x\rfloor+1} W^{(n,i)} > g(\lfloor m^n x\rfloor + 1)\right) \geq p \text{ (by (8.36))}. \tag{8.37}$

This observation connects the right tails of the distributions of W with that of $\sup_{n\geq 1} W_n$ as

$$\begin{aligned} pP(\sup_{n\geq 1} W_n > x) &= p \sum_{n\geq 1} P(\mathsf{F}_n(x)) \\ &\leq \sum_{n\geq 1} P(W > gx|\mathsf{F}_n(x))P(\mathsf{F}_n(x)) \text{ (by (8.37))} \\ &= P(W > gx). \end{aligned}$$

Since this holds for any $g \in (0, \mathrm{E}(W))$ and $x > 0$, we conclude that

$$\mathrm{E}(\sup_{n\geq 1} W_n) < \infty.$$

Hence by Martingale convergence Theorem 5.2.3, $\lim_{n\to\infty} W_n$ exists a.s. So we have

$$\begin{aligned} W &= W_0 + \sum_{n=0}^{\infty}(W_{n+1} - W_n) \\ &= 1 + \sum_{n=0}^{\infty}(W_{n+1} - W_n), \end{aligned}$$

where

$$W_{n+1} - W_n = m^{-(n+1)} \sum_{i=1}^{Z_n} (\xi_{n,:i} - m). \tag{8.38}$$

Define

$$\begin{aligned} \widetilde{W}_{n+1} &:= m^{-(n+1)} \sum_{i=1}^{Z_n} \xi_{n,i} \mathbf{1}_{\{\xi_{n,i}\leq m^n\}} \\ R_n &:= \mathrm{E}(W_{n+1} - \widetilde{W}_{n+1}|\mathcal{G}_n) = \mathrm{E}(W_n - \widetilde{W}_{n+1}|\mathcal{G}_n) \\ &= m^{-(n+1)} \sum_{i=1}^{Z_n} \mathrm{E}(\xi_{n,i} \mathbf{1}_{\{\xi_{n,i}>m^n\}}|\mathcal{G}_n) \\ &= m^{-1} W_n \mathrm{E}(\xi \mathbf{1}_{\{\xi>m^n\}}). \end{aligned}$$

In the next lemmas we approximate the series in (8.38) by the series with the terms $(\widetilde{W}_{n+1} - W_n + R_n : n \geq 0)$. Their proofs are given later.

Lemma 8.9.2 *If* $1 < \mathrm{E}(\xi) < \infty$, *then* $(\widetilde{W}_n : n \geq 0)$ *and* $(W_n : n \geq 0)$ *are tail equivalent. That is,*

$$P(\widetilde{W}_n \neq W_n \text{ infinitely often}) = 0.$$

♦

8.9 Branching Process: Kesten–Stigum Theorem

Lemma 8.9.3 *If $1 < \mathrm{E}(\xi) < \infty$, then $\sum_{n=0}^{\infty} \mathrm{Var}(\widetilde{W}_{n+1} - W_n + R_n) < \infty$. As a consequence, $\sum_{n=1}^{\infty}(\widetilde{W}_{n+1} - W_n + R_n)$ converges in L^1 and almost surely.* ♦

Exercise 8.9.2 If $1 < \mathrm{E}(\xi) < \infty$, then $\sum_{n=0}^{\infty} R_n$ converges in L^1 if and only if Condition (1) of Theorem 8.9.1 holds.

Step 2: Condition (1) $\Rightarrow$ Condition (3). Lemma 8.9.2 and Exercise 8.9.2 imply that $\sum_{n=1}^{\infty}(\widetilde{W}_{n+1} - W_n)$ converges a.s. and in L^1. Therefore,

$$\lim_{N \to \infty} \mathrm{E}\left(\sum_{i=N}^{\infty}(\widetilde{W}_{i+1} - W_i)\right) = 0. \tag{8.39}$$

This implies that for any $n \geq 1$,

$$\mathrm{E}(W) = \mathrm{E}\left(1 + \sum_{n=0}^{N}(W_{n+1} - W_n) + \sum_{n=N+1}^{\infty}(W_{n+1} - W_n)\right)$$

$$\geq 1 + \mathrm{E}\sum_{i=0}^{N}(W_{n+1} - W_n) + \mathrm{E}\left(\sum_{n=N+1}^{\infty}(\widetilde{W}_{n+1} - W_n)\right). \tag{8.40}$$

The second sum is 0 for every N. As $N \to \infty$, the last term converges to 0 by (8.39). On the other hand, by Fatou's lemma,

$$\mathrm{E}(W) \leq \liminf_{n \to \infty} \mathrm{E}(W_n) = 1.$$

Hence, we conclude $\mathrm{E}(W) = 1$.

Step 3: Condition (2) $\Rightarrow$ Condition (1). Let $W_* = \inf_{n \geq 1} W_n$.

Note that
$$P(W_* > 0) = P(W > 0) > 0.$$

It follows from Lemma 8.9.3 and 8.9.2 that $\sum_{n=1}^{\infty}(W_{n+1} - W_n + R_n)$ converges almost surely, and so $\sum_{n=1}^{\infty} R_n$ must converge almost surely.

Now observe that

$$\sum_{n=0}^{\infty} R_n = m^{-1} \sum_{n=0}^{\infty} W_n \mathrm{E}(\xi \mathbf{1}_{\{\xi > m^n\}})$$

$$\geq W_* m^{-1} \sum_{n=0}^{\infty} \mathrm{E}(\xi \mathbf{1}_{\{\xi > m^n\}}) \text{ almost surely.}$$

This implies that $\sum_{n=0}^{\infty} m^n \mathrm{E}(\xi \mathbf{1}_{\{\xi > m^n\}}) < \infty$.

For some constant C,

$$\infty > \sum_{n=0}^{\infty} E(\xi \mathbf{1}_{\{\xi > m^n\}}) = E\Big(\xi \sum_{n=0}^{\infty} \mathbf{1}_{\{\xi > m^n\}}\Big)$$
$$= E\Big(\xi \sum_{n=0}^{\infty} \mathbf{1}_{\{n < (\log m)^{-1} \log_+ \xi\}}\Big) \qquad (8.41)$$
$$\geq CE(\xi \log_+ \xi) \text{ (by Fubini's theorem)}.$$

Hence, the proof follows. □

Proof of Lemma 8.9.2 By Borel–Cantelli lemma, it is enough to show $\sum_{n=0}^{\infty} P\big(\widetilde{W}_{n+1} \neq W_{n+1}\big) < \infty$. Observe that the left side equals

$$\sum_{n=0}^{\infty} P\Big(\sum_{i=1}^{Z_n} \mathbf{1}_{\{\xi_{n,i} > m^n\}} \geq 1\Big) \leq \sum_{n=0}^{\infty} E\Big(\sum_{i=1}^{Z_n} \mathbf{1}_{\{\xi_{n,i} > m^n\}}\Big)$$
$$= \sum_{n=0}^{\infty} E(Z_n) P(\xi > m^n)$$
$$= \sum_{n=0}^{\infty} m^n E(\mathbf{1}_{\{\xi > m^n\}}) = E\Big(\sum_{n=0}^{\infty} m^n \mathbf{1}_{\{\xi > m^n\}}\Big).$$

It is easy to see that

$$\sum_{n=0}^{\infty} m^n \mathbf{1}_{\{n < (\log_+ \xi)/(\log m)\}} \leq (m-1)^{-1} \max(\xi, 1) \text{ almost surely as } m > 1.$$

But $E(\max(\xi, 1)) < \infty$, and hence the lemma follows. □

Proof of Lemma 8.9.3 The variables being centered (given Z_n),

$$\sum_{n=0}^{\infty} \text{Var}\big(\widetilde{W}_{n+1} - W_n + R_n\big)$$
$$= \sum_{n=0}^{\infty} E\Big(\text{Var}\big(\widetilde{W}_{n+1} - W_n + R_n \mid Z_n\big)\Big)$$
$$= \sum_{n=0}^{\infty} E\Big(\text{Var}\Big(\frac{1}{m^{n+1}} \sum_{i=1}^{Z_n} \big(\xi_{n,i} \mathbf{1}_{\{\xi_{n,i} \leq m^n\}} - mE(\xi \mathbf{1}_{\xi \leq m^n})\big) \mid Z_n\Big)\Big)$$

$$= \sum_{n=0}^{\infty} \frac{\mathrm{E}(Z_n)}{m^{2n+2}} \mathrm{Var}\Big(\xi \mathbf{1}_{\{\xi \le m^n\}} - m\mathrm{E}\big(\xi \mathbf{1}_{\xi \le m^n}\big)\Big)$$

$$\le m^{-2} \sum_{n=0}^{\infty} m^{-n} \mathrm{E}(\xi^2 \mathbf{1}_{\{\xi \le m^n\}})$$

$$= m^{-2} \mathrm{E}\Big(\xi^2 \sum_{n=0}^{\infty} m^{-n} \mathbf{1}_{\{n \ge (\log_+ \xi)/(\log m)\}}\Big)$$

$$\le \frac{1}{m(m-1)} \mathrm{E}\Big(\frac{\xi^2}{\max(\xi, 1)}\Big) < \infty.$$

Hence, using Borel–Cantelli lemma, the proof is complete. □

References

S. Asmussen, H. Hering, *Branching Processes*, vol. 3. (Springer, 1983)

K.B. Athreya, P.E. Ney, *Branching Processes*. (Dover Publications, 2004)

B. Bercu, B. Delyon, E. Rio, *Concentration Inequalities for Sums and Martingales*. (Springer International Publishing, 2015)

S. Bernstein, On a modification of Chebyshev's inequality and of the error formula of Laplace. Ann. Sci. Inst. Sav. Ukraine, Sect. Math (1924)

S. Bernstein, *Theory of Probability* (In Russian) (Moscow, 1927)

S. Bernstein, On certain modifications of Chebyshev's inequality. Doklady Akademii Nauk SSSR **17**(6), 275–277 (1937)

B.M. Brown, A general three series theorem. Proc. Am. Math. Soc. **28**(2), 573–577 (1971)

D. L. Burkholder and R. F. Gundy. Extrapolation and interpolation of quasi-linear operators on martingales. *Acta Mathematica*, **124**(none), 249 – 304 (1970)

Y. S. Chow, H. Teicher, *Probability Theory: Independence, Interchangeability, Martingales*. (Springer Science & Business Media, 2003)

B. Davis, On the intergrability of the martingale square function. Israel J. Math. **8**, 187–190 (1970)

P. Hall, C.C. Heyde, *Martingale Limit Theory and its Application* (Academic, Harcourt Brace Jovanovich, Publishers, New York-London, 1980)

W. Hoeffding, On sequences of sums of independent random vectors, in *Proceedings of 4th Berkeley Symposium Mathematical Statistics and Probability*, Vol. II (University of California Press, 1961), pp. 213–226

H. Kesten, B.P. Stigum, A limit theorem for multidimensional Galton-Watson processes. Ann. Math. Stat. **37**(5), 1211–1223 (1966)

E. Lenglart, Semi-martingales et intégrales stochastiques en temps continu. Revue du CETHEDEC **20**(75), 91–160 (1983)

M. Loève, *Elementary Probability Theory*. (Springer, 1977)

C. McDiarmid, On the method of bounded differences. Surv. Combin. **141**(1), 148–188 (1989)

D. Revuz, M. Yor, *Continuous Martingales and Brownian Motion*, vol. 293. (Springer Science & Business Media, 2013)

Index

Symbols
0-1 law, Hewitt–Savage, 103
0-1 law, Kolmogorov, 97
L^1, 88
L^2, Cauchy, 77
L^p bounded martingale, 89
L^p space, metric, 7
U-statistics, CLT, 143
U-statistics, SLLN, 101
U-statistics, degree, order, 100
U-statistics, forward martingale, 141
U-statistics, kernel, 100
σ-algebra, 2
σ-field, 2
σ-field, Borel, 2
σ-field, complete, 4
σ-field, completion, 68
σ-field, independent, 102
σ-field, stopped, 69
σ-field, tail, 97, 106
σ-field, tail invariant, 103
σ-field, trivial, 4
σ-finite measure, 3, 5

A
Abracadabra, 80
Absolute continuity, measures, 27
Absolutely continuous function, 34
Adapted, 61, 69
Algebra, 1
Almost sure convergence, 6
Almost surely, 3
Approximation theorem, 12
Azuma-Hoeffding inequality, strong, 146
Azuma-Hoeffding inequality, weak, 143

B
Bernoulli distribution, 15, 116
Bernoulli random variable, 105
BGW process, 65, 167
Binomial distribution, 15
Borel–Cantelli Lemma, 4
Borel σ-field, 2
Bounded stopping time, 68
Bound, polynomial, 143
Branching process, Galton–Watson, 65
Burkholder-Davis-Gundy inequality, 158

C
Cantor function, 38
Cantor set, 37
Carathéodory's extension theorem, 4
Cauchy in L^2, 77
Cauchy–Schwarz inequality, 7
Cauchy–Schwarz inequality, conditional, 51, 52
Central Limit Theorem, 79, 119
Chain, irreducible, 94
Chain, recurrent, 94
Characteristic function, 17
Chebyshev's inequality, 7
Chromatic number, 145
CLT, 119
CLT for U-statistics, 143
CLT, Lévy–Lindeberg, 119
CLT, martingale, 120
Compact set, 12, 13, 51
Compensator, 62

Complete σ-field, 4
Completion, σ-field, 68
Conditional Cauchy–Schwarz inequality, 51, 52
Conditional distribution function, 109
Conditional expectation, 41
Conditional expectation, discrete, 43
Conditional three series theorem, 151
Conditional variance, 121
Continuity of measure, 24
Continuity theorem, Lévy, 17
Convergence, almost sure, 6
Convergence, in distribution, 16
Convergence, in measure, 6
Convergence, in probability, 6, 16
Convergence, reverse martingale, 91
Convex, 51
Countably generated, 46
Coupling, 137, 138
Covariance, 7
Cumulative Distribution Function (CDF), 16

D
DCT, 6, 45, 76–78
DCT vs. UI, 53
DCT, in probability, 6
Decomposition, Doob, 62
Decomposition, Jordan–Hahn, 25
de-Finetti's theorem, 110
Discontinuity points, 20
Discrete harmonic function, 67
Distribution, Bernoulli, 15, 116
Distribution, binomial, 15
Distribution convergence, 16
Distribution, exponential, 16
Distribution function, 8
Distribution function, conditional, 109
Distribution, gamma, 16
Distribution, Gaussian, 15, 79
Distribution, Poisson, 15, 64
Distribution, uniform, 58, 116
Dominated convergence theorem, 6, 72, 73
Doob decomposition, 62, 84
Doob's martingale, 65, 88, 89
Doob's maximal inequality, 79, 80, 90, 91
Doob's optional stopping theorem, 71, 72, 74, 79, 86, 150
Doubling game, 64, 88
Downward theorem, Lévy, 92

E
Erdős-Rényi graph, 145

Exchangeable, 105
Expectation, 5
Expectation, conditional, 41
Exponential distribution, 16
Exponential inequality, 143
Exponential inequality, martingale, 143
Extended Borel-Cantelli lemma, 150
Extended real numbers, 2
Extinction, 168

F
Fatou's lemma, 6, 76, 87
Field, 1
Field, smallest, 99
Filtration, 61
Filtration, reverse, 91
Finitely additive, 3
Finite measure, 3
Finite stopping time, 68
Forward martingale, U-statistics, 141
Fubini's theorem, 5
Function, absolutely continuous, 34
Function, measurable, 4
Function, positive definite, 21
Function, simple, 4, 46, 49

G
Galton–Watson branching process, 65
Gamma distribution, 16
Gaussian distribution, 15, 79
Gaussian mixture, 120
Gaussian, multivariate, 16
Gaussian random variable, 105
Generator, countable, 46
Graph, Erdős-Rényi, 145

H
Harmonic function, discrete, 67
Heine–Borel theorem, 11
Hewitt-Savage 0-1 law, 103
Hitting time, 68
Hölder's inequality, 7, 90
Hölder's inequality, conditional, 52

I
Identity, Wald, 76
IID, 18
IID, strong law, 98
Indefinite integral, 23
Indefinite integral, Jordan–Hahn, 27
Independent σ-fields, 102

Index 177

Independent random variables, 18
Induced measure, 14
Inequality, Azuma-Hoeffding strong, 146
Inequality, Azuma-Hoeffding weak, 143
Inequality, Burkholder-Davis-Gundy, 158
Inequality, Cauchy–Schwarz, 7, 51, 52
Inequality, Chebyshev, 7
Inequality, conditional Jensen's, 50
Inequality, exponential, 143
Inequality, Hölder, 7, 52
Inequality, Lyapunov, 7
Inequality, Markov, 7, 78
In probability, DCT, 6
Integral, indefinite, 23
Invariant σ-field, 103
Irreducible, 94

J
Jensen's inequality, 146
Jensen's inequality, conditional, 50
Jordan–Hahn decomposition, 25, 34
Jordan–Hahn, indefinite integral, 27
Jordan–Hahn, minimality, 36

K
Kesten-Stigum theorem, 168
Kolmogorov's 0-1 law, 97, 100, 103
Kolmogorov's maximal inequality, 78
Kronecker's lemma, 153, 156

L
Lévy continuity theorem, 17
Lévy–Lindeberg CLT, 119
Lévy's downward theorem, 92
Lévy's theorem, 92
Lévy's upward theorem, 88, 98, 104, 106
Law of large numbers, weak, 19
Lebesgue decomposition theorem, 33
Lebesgue measure, 12
Lebesgue–Stieltjes measure, 8
Lemma, Borel–Cantelli, 4, 150
Lemma, Fatou, 6, 87
Lemma, Kronecker, 153
Lemma, Toeplitz, 152
Lemma, upcrossing, 86, 93
Likelihood ratio, 113

M
Markov chain, 66, 94, 133
Markov inequality, 7, 19, 55, 78

Martingale, 72–74, 78
Martingale CLT, 120, 125
Martingale convergence theorem, 121, 150, 151
Martingale, definition, 62
Martingale, Doob's, 89
Martingale, exponential inequality, 143
Martingale, L^p bounded, 89
Martingale, product, 64
Martingale, reverse, 91
Martingale, SLLN, 154
Martingale, stopped, 70
Martingale, uniformly integrable, 88
Martingale weak law, 148
Maximal element, 29
Maximal inequality, Doob's, 79, 91
Maximal inequality, Kolmogorov, 78
Mean, 5
Measurable function, 4
Measurable space, 2
Measure, 3
Measure, σ-finite, 3, 5
Measure, absolutely continuity, 27
Measure, continuity, 24
Measure, finite, 3
Measure, induced, 14
Measure, Lebesgue, 12
Measure, Lebesgue–Stieltjes, 8
Measure, probability, 3
Measure, product, 12
Measure, signed, 23
Measure, total variation, 27
Metric space, L^p, 7
Mixing, 156
Mixture, Gaussian, 120
Monotone class theorem, 2, 13, 89, 99
Monotone Convergence Theorem, 6, 76, 87
Multivariate Gaussian distribution, 16

N
Null set, 3

O
Open set, 2, 12, 14, 35, 37
Optional Stopping Theorem (OST), 71, 72, 74, 79, 86, 150

P
Partial sum, 19
Permutation, 101, 104
Poisson distribution, 15, 64

Poisson process, 64
Pólya's urn, 65, 91
Polynomial bound, 143
Positive definite function, 21
Probability density, 15
Probability distribution, 14
Probability distribution, family, 58
Probability space, filtered, 61
Product martingale, 64
Product measure, 12

Q
Quadratic variation, 84, 94, 125, 158, 165

R
Radon–Nikodym theorem, 28
Radon-Nikodym derivative, 93
Random matrix, 129
Random variable, 4
Random variable, Bernoulli, 105
Random variable, Gaussian, 105
Random variable, independence, 18
Random walk, simple symmetric, 64, 73
Rao-Blackwell theorem, 58
Real numbers, extended, 2
Recurrence, 94
Regular Conditional Distribution (RCD), 48, 148
Reverse filtration, 91, 99
Reverse martingale, 91
Reverse martingale convergence, 91, 102, 106, 109, 110
Ruin probability, 73, 74

S
Sample mean, 101
Sample variance, 101
Semi-field, 1, 98
Set, compact, 12, 13, 51
Set, open, 2, 12, 14, 35, 37
Signed measure, 23
Simple function, 4, 46, 49
Simple symmetric random walk, 64, 73
Singular, 32
SLLN for martingales, 154
Smallest field, 99
Space, measurable, 2
State space, 66, 94
Stopped martingale, 70
Stopped process, 69
Stopped σ-field, 69

Stopping time, 68
Stopping time, finite, bounded, 68
Strong Azuma-Hoeffding inequality, 146
Strong law, 60
Strong law for U-statistics, 101
Strong law for iid, 98
Strong Law of Large Numbers (SLLN), 60, 99
Sub-martingale, 63, 64, 78, 90, 91
Sub-martingale, definition, 62
Sub-martingale, L^1-bounded, 87
Sub-martingale, maximal inequality, 79
Sub-martingale, OST, 71
Sub-martingale, uniformly integrable, 88
Sub-martingale, upcrossing, 86, 93
Sufficient statistic, 58
Super-martingale, definition, 62
Survival event, 168

T
Tail σ-field, 97, 106
Theorem, approximation, 12
Theorem, Carathéodory, 4
Theorem, de-Finetti, 110
Theorem, dominated convergence, 6
Theorem, Fubini, 5
Theorem, Kesten-Stigum, 168
Theorem, Lebesgue decomposition, 33
Theorem, Lévy, 92
Theorem, monotone class, 2
Theorem, monotone convergence, 6
Theorem, Radon–Nikodym, 28
Theorem, Rao-Blackwell, 58
Three series theorem, conditional, 151
Tight, 79
Toeplitz lemma, 152
Total variation measure, 27
Tower property, 44, 142
Trace, 129
Transition probability, 66, 94
Trivial σ-field, 4, 100

U
UI vs. DCT, 53
Unbiased estimator, 58
Uniform distribution, 58, 116
Uniformly Integrable (UI), 52–54, 59, 72, 88–90
Upcrossing, 85, 92
Upcrossing lemma, 86, 87
Upcrossing lemma, general, 93
Upward theorem, Lévy, 88, 98, 104, 106

Index

Urn models, 126
Urn, Pólya, 65

V
Variance, 7
Variance, conditional, 121
Variation, quadratic, 84

W
Wald's first identity, 76
Wald's second identity, 76
Wald's third identity, 77
Weak Azuma-Hoeffding inequality, 143
Weak convergence, 16
Weak law of large numbers, 19
Weak law, martingale, 148
Wigner matrix, 129

Author Index

A
Ash, R., vii, 1
Asmussen, S., 168
Athreya, K. B., 168

B
Bercu, B., 143
Berk, R., 101
Bernstein, S., 143
Billingsley, P., vii, 1, 12, 60
Bingham, M. S., 21
Blackwell, D., 94, 95
Breiman, L., vii, 1
Brown, B. M., 151
Burkholder, D. L., 157

C
Caravenna, F., 80
Chow, Y. S., vii, 157, 158
Chung, K. L., vii

D
Davis, B., 157
De Finetti, B., 107
Delyon, B., 143
Doléans-Dade, C. A., vii, 1
Doob, J. L., vii, 61
Dubins, L., 94, 95
Durrett, R., vii

E
Elekes, M., 38

F
Freedman, D. A., 126
Friedman, B., 126

G
Gundy, R. F., 157

H
Hall, P., vii, 84, 120
Hering, H., 168
Hewitt, E., 103
Heyde, C. C., vii, 84, 120
Hoeffding, W., 141

K
Keleti, T., 38
Kesten, H., 168
Kirsch, W., 107

L
Lenglart, E., 158
Lévy, P., 61, 119
Lindeberg, J. W., 119
Lukacs, E., 17

M
McDiarmid, C., 143

N
Neveu, J., vii
Ney, P. E., 168
Norris, J. R., 94, 133, 134, 136, 138

P
Parthasarathy, K. R., 21

R
Ramachandran, B., 17
Revuz, D., 158
Rio, E., 143
Robbins, H., 95

S
Savage, L. J., 103
Siegmund, D., 95
Stigum, B. P., 168

T
Teicher, H., vii, 157

V
Ville, J., 61

W
Wald, A., 76
Williams, D., vii, 80

Y
Yor, M., 158

SPRINGER NATURE

GPSR Compliance

The European Union's (EU) General Product Safety Regulation (GPSR) is a set of rules that requires consumer products to be safe and our obligations to ensure this.

If you have any concerns about our products, you can contact us on ProductSafety@springernature.com

In case Publisher is established outside the EU, the EU authorized representative is:

Springer Nature Customer Service Center GmbH
Europaplatz 3
69115 Heidelberg, Germany

The manufacturer's authorised representative in the EU is Springer Nature Customer Service Centre GmbH, Europaplatz 3, 69115 Heidelberg, Germany. If you have any concerns regarding our products, please contact ProductSafety@springernature.com

Printed and bound by CPI Group (UK) Ltd, Croydon, CR0 4YY

26/03/2026

02078966-0001